History of Mechanism and Machine Science

Volume 32

Series editor

Marco Ceccarelli
LARM: Laboratory of Robotics and Mechatronics
DICeM; University of Cassino and South Latium
Via Di Biasio 43, 03043 Cassino (Fr), Italy
ceccarelli@unicas.it

Aims and Scope of the Series

This book series aims to establish a well defined forum for Monographs and Proceedings on the History of Mechanism and Machine Science (MMS). The series publishes works that give an overview of the historical developments, from the earliest times up to and including the recent past, of MMS in all its technical aspects.

This technical approach is an essential characteristic of the series. By discussing technical details and formulations and even reformulating those in terms of modern formalisms the possibility is created not only to track the historical technical developments but also to use past experiences in technical teaching and research today. In order to do so, the emphasis must be on technical aspects rather than a purely historical focus, although the latter has its place too.

Furthermore, the series will consider the republication of out-of-print older works with English translation and comments.

The book series is intended to collect technical views on historical developments of the broad field of MMS in a unique frame that can be seen in its totality as an Encyclopaedia of the History of MMS but with the additional purpose of archiving and teaching the History of MMS. Therefore the book series is intended not only for researchers of the History of Engineering but also for professionals and students who are interested in obtaining a clear perspective of the past for their future technical works. The books will be written in general by engineers but not only for engineers.

Prospective authors and editors can contact the series editor, Professor M. Ceccarelli, about future publications within the series at:

LARM: Laboratory of Robotics and Mechatronics
DICeM; University of Cassino and South Latium
Via Di Biasio 43, 03043 Cassino (Fr)
Italy
email: ceccarelli@unicas.it

More information about this series at http://www.springer.com/series/7481

Carlos López-Cajún · Marco Ceccarelli
Editors

Explorations in the History of Machines and Mechanisms

Proceedings of the Fifth IFToMM Symposium on the History of Machines and Mechanisms

 Springer

Editors
Carlos López-Cajún
Department of Mechanical Engineering
Universidad Autónoma de Querétaro
Santiago de Querétaro
Mexico

Marco Ceccarelli
LARM: Laboratory of Robotics and
  Mechatronics
DICeM; University of Cassino and
  South Latium
Cassino
Italy

ISSN 1875-3442　　　　　　　ISSN 1875-3426　(electronic)
History of Mechanism and Machine Science
ISBN 978-3-319-31182-1　　　ISBN 978-3-319-31184-5　(eBook)
DOI 10.1007/978-3-319-31184-5

Library of Congress Control Number: 2016934426

© Springer International Publishing Switzerland 2016
This work is subject to copyright. All rights are reserved by the Publisher, whether the whole or part of the material is concerned, specifically the rights of translation, reprinting, reuse of illustrations, recitation, broadcasting, reproduction on microfilms or in any other physical way, and transmission or information storage and retrieval, electronic adaptation, computer software, or by similar or dissimilar methodology now known or hereafter developed.
The use of general descriptive names, registered names, trademarks, service marks, etc. in this publication does not imply, even in the absence of a specific statement, that such names are exempt from the relevant protective laws and regulations and therefore free for general use.
The publisher, the authors and the editors are safe to assume that the advice and information in this book are believed to be true and accurate at the date of publication. Neither the publisher nor the authors or the editors give a warranty, express or implied, with respect to the material contained herein or for any errors or omissions that may have been made.

Printed on acid-free paper

This Springer imprint is published by Springer Nature
The registered company is Springer International Publishing AG Switzerland

Preface

The organization of an international symposium on the History of Machines and Mechanisms (HMM) every 4 years is the main activity of the Permanent Commission (PC) for the History of Mechanism and Machine Science of IFToMM, the International Federation for the Promotion of Mechanism and Machine Science. The first two symposia, HMM2000 and HMM2004 were held at the University of Cassino in Cassino, Italy in the years 2000 and 2004. The third symposium, HMM2008, was held at the National Cheng Kung University in Tainan, Taiwan in 2008. The fourth symposium was held at Vrije University in Amsterdam, The Netherlands in 2012. The present volume contains the proceedings of HMM2016, the fifth International Symposium on the History of Machines and Mechanisms that was held at Queretaro University in Queretaro, Mexico, from June 7 to 9, 2016.

The mission of IFToMM is to promote research and development in the field of machines and mechanisms by theoretical and experimental methods, along with their practical applications. The aim of the international symposia on HMM is to maintain an international forum for the exploration of the history of machines and mechanism. Since the emphasis is on the history of technical systems and their applications, the scope of the symposia is wide. Relevant topics are also the history of theories and design methods, biographies, the history of the institutions involved, the relations with other disciplines, the history of engineering education, and the social and cultural aspects of machines.

History is not simply full of exciting and entertaining stories. Historical investigations put our own present-day activities in a wider perspective. They help us define who we are. Moreover history remains a source of ideas.

This book is meant for researchers, graduate students, engineers, and all others with an interest in the history of machines and mechanisms. We believe that it can inspire and motivate them.

After the review process 23 papers by authors representing ten different countries were accepted for publication in the proceedings of HMM2016. One glance at the table of contents is enough to see that we succeeded in bringing together an interesting group of people with a stimulating variation in subjects. We are very satisfied with this result and we thank the authors for their valuable contributions and for the efforts in submitting in time the final versions of the papers.

We would like to express our sincere gratitude to the members of the scientific committee:

M. Ceccarelli, University of Cassino, Italy, Chair
T. Chondros, University of Patras, Greece
O. Egorova, Bauman Moscow State Technical University (BMSTU), Russia
H. Kerle, Technical University of Braunschweig, Germany
T. Koetsier, Vrije University, The Netherlands
C. Lopez-Cajún, University of Querétaro, México
J.S. Rao, India Institute of Technology, India
L. Zhen, Beijing University, Beijing, China
H.S. Yan, Cheng Kung University, Taiwan, Taipei, China
B. Zhang, Chinese Academy of Science, Beijing, China

Moreover, we would also like to thank the colleagues who helped us in the review process.

We also thank the sponsors of the symposium: IFToMM, The Department of Mechanical Engineering of the School of Engineering of the Universidad Autónoma de Querétaro. Many thanks to Professors Gilberto Herrera, Aurelio Domínguez and Juan Primo Benítez. Without their support we would not have been able to organize HMM2016.

January 2016

Carlos López-Cajún
Marco Ceccarelli

Contents

Aspects of the Cost-Effectiveness of Restoration Process of the F. Reuleaux Mechanisms 1
D. Spasskaya and N. Terehova

Application of Rapid Prototyping Technology for Modeling the Mechanisms of F. Reuleaux Collection 9
M. Vlasov and M. Samoylova

Protoepistemology of Mechanical Engineering in Cassiodorus' *Variae* or Mission Impossible at Theoderic's Court 17
N. Ambrosetti

The Silk Mill *"alla Bolognese"* 31
C. Bartolomei and A. Ippolito

Robots in History: Legends and Prototypes from Ancient Times to the Industrial Revolution 39
A. Gasparetto

The First Hundred Years of Mechanism Science at RWTH Aachen University 51
B. Corves

Mock-Up of an Eighteenth-Century Oil Mill via Rapid-Prototyping 65
J.C. Montes, R. López-García, R. Dorado-Vicente and F.J. Trujillo

An Introduction to the Ancient Mechanical Wind-Instrument Automata 77
Yu-Hsun Chen, Jian-Liang Lin and Hong-Sen Yan

Dynamic Reconstruction of a Colonial Mexican Mechanism 87
J.C. Jauregui-Correa and G. Rodriguez-Zahar

Leibniz's Developments of Machine Science 101
A.R.E. Oliveira

**On the Mechanics of Live Nature in the Works of
V.P. Goryachkin** 113
V. Chinenova

**The Transformation of the Largest Aircraft Factory of Romania
in Tractors Factory as Result of the Soviet Occupation** 125
H. Salcă and D. Săvescu

**New Trends in Learning Through 3D Modeling of Historical
Mechanism's Model** 137
O. Egorova, K. Samsonov and A. Sevryukova

Some Inventions by Engineers of the Hellenistic Age 151
C. Rossi

Mechanical Engineer DING Gongchen 165
Zhang Baichun and Liu Yexin

Lewis Mumford Revisited 171
Teun Koetsier

**Analysis of Structure, Kinematic and 3D Modeling of Ferguson's
Mechanisms** 183
V. Tarabarin and A. Kozov

**19th c. Olivier String Models at Cornell University: Ruled Surfaces
in Gear Design** 195
F.C. Moon and J.F. Abel

**Mechanism of Laoguanshan Pattern Looms from Late 2nd Century
BCE, Chengdu, China** 209
Feng Zhao, Yi Wang, Qun Luo, Bo Long, Baichun Zhang, Yingchong Xia
and Tao Xie

On the Warship by Ansaldo for Chinese Imperial Navy 223
Yibing Fang and Marco Ceccarelli

Dynamic Analysis of an Ancient Tilt-Hammer 235
Umberto Meneghetti

**An Analysis of Micro Scratches on Typical Southern Chinese
Bronzes—A Case Study of Crawler-Pattern Nao and Chugong Ge** 245
Lie Sun, Xiaowu Guan and Shilei Wu

Aspects of the Cost-Effectiveness of Restoration Process of the F. Reuleaux Mechanisms

D. Spasskaya and N. Terehova

Abstract Models from the collection of the F. Reuleaux mechanisms stored in Bauman Moscow State Technical University (BMSTU) not only belong to the objects of cultural heritage but also are actively being used for the educational purposes. However, this leads to a gradual deterioration of the working properties of the components of the mechanisms that is why it is necessary to carry out the restoration process. Improvement of the technologies used during the restoration process increases the durability of mechanisms parts but at the same time greatly increases the costs of their implementation. The assessment of expenses for restoration of model Q-17 from the F. Reuleaux collection has shown that it is possible to pick up the best materials and technologies for carrying out the restoration process as well as has shown the expediency and a possibility of manufacturing of prototypes of models of mechanism Q-17 with use of technology of the rapid prototyping realized with the use of the three-dimensional printing system Objet Eden 250.

Keywords F. Reuleaux · Rapid prototyping · Aspects of the cost-effectiveness of restoration process · Theory of mechanisms and machines · Restoration · 3D-modeling · Three-dimensional printing

1 Introduction

The Models of the mechanisms from F. Reuleaux collection stored in the BMSTU belong to the objects of scientific, technical and cultural heritage. The model of the mechanism Q-17 that is shown in the Fig. 1 is used in the educational process at Bauman University. However, it leads to the gradual deterioration of working

---

D. Spasskaya (✉) · N. Terehova
The Bauman Moscow State Technical University, Moscow, Russia
e-mail: spasskaydd@mail.ru

N. Terehova
e-mail: terehova_n_u@mail.ru

Fig. 1 Model S8 of the mechanisms from the Franz Reuleaux collection (Q-17)

properties of the model of the mechanism. It is possible to extend the «life» of such unique models by conducting the process of restoration.

Restoration process consists of a number of the different actions providing the conservation and disclosing of a historical and scientific heritage of the models of the mechanisms. Repair of the unique models of the mechanisms, including periodic repair work, represents the set of some actions taken to maintain their technical without changing the existing shape of the mechanism. Adaptation is also included in the set of the repair activities and it is aimed at the creating the conditions for a modern use of historical models of mechanism without compromising its historical and scientific value. Thus, restoration, repair and adaptation of the models of the mechanisms can be carried out. Moreover, these activities can be done in the different combinations, which is important for a collection of the mechanisms of Franz Reuleaux collection [1].

In some cases, the term 'restoration' applies only to the separate elements and details of an object that should be saved. As for their whole object, there are always elements, parts, items requiring 'reconstruction'. Therefore, some of the work and projects of restoration is necessary to specify, for example, the 'restoration' project with elements of 'reconstruction'. In practice, this concept is often replaced by the term 'recreation' Thus, the recreation is a set of measures to restore the lost that seems possible only if there is enough scientific and technical data about an object. All types of works with models: restoration, repair, reconstruction, adaptation and recreation considered to be a single process aimed to maintain the models in a working condition and accompanied by the industrial, research and project work. While identifying an intensive physical wear of the models of the mechanisms, essential measures to restore their technical properties are required.

Before the beginning of the restoration process, it is necessary to analyze the reasons for internal and external damages of models as the method and technology for the restoration works depend on them.

Repair and restoration works of the model of the mechanism of F. Reuleaux collection carried out by structural departments of the Bauman University. The works are carried out in compliance with all requirements and regulations for the restoration of objects of scientific and historical heritage.

Design work on restoration and repair must be carried out in accordance with the modern rules and regulations as well as accompanied by the development of scientific and project documentation, which represents a set of research, design, surveying and alignment documentation that includes:

- preliminary work;
- integrated research;
- restoration job;
- scientific explanation of design decisions;
- the basic principles on the organization of work;
- financial estimates;
- estimates made by analyzing the working drawings;
- scientific and restoration report;
- scientific substantiation of design decisions with the preliminary design drawings;
- drawings and three-dimensional computer models;
- research and restoration report

The list of restoration reports, shown above, includes the recommendations for all types of research, survey, design and production work that should be done and it is aimed at research and preservation of models of F. Reuleux mechanisms that have value from the point of history, art, science and technology [2].

It is required that at all the stages of the modern scientific restoration process all technical and technological information contained in the mechanisms must be saved, that is why the used techniques and the restoration materials should not distort this information.

Preparation of scientific documentation for restoration, preservation and related studies is accompanied by the photo and video materials.

Modern computer programmes are used to conduct a three-dimensional computer modelling of the mechanisms. Carrying out the analysis of the state of the mechanisms and identified defects is performed with the use of specialized computer software [3, 4].

The main purpose of the preliminary survey is to assess the state of the model of the mechanism as well as to obtain the necessary information for the compilation of a detailed programme of instrumental studies.

Based on the results of the preliminary analysis of the engineering research causes of defects and damages are identified, but during the following stages of restoration process some adjustments to the data are possible.

The restoration task for the research of the models of the F. Reuleaux mechanisms has been developed based on the materials of the preliminary works and contains the data about a necessity of the development of the several versions of the project as well as the suggestions on the sequence of carrying out the different stages of the restoration process made by months.

While doing a preliminary survey of the models of the mechanisms the following parameters should be established: the type of constructions, their constructive diagrams and types of the connection of the elements; the existence and nature of the deformities and damage structures, the connections of the elements and nodes; physical-mechanical characteristics (Fig. 2).

During the restoration process, issues related to the dismantling of the model of the mechanism Q-17 have been solved. Figure 3 shows the elements of the model of the mechanism Q-17. The main elements of the model are considered to be rack mechanism, crank, connecting rods and the slide. Dismantling allowed us to make the measurement process and provide a three-dimensional computer modeling and three-dimensional printing of the prototype with the use of 3D printing system Eden 250.

During the reconstruction process of historical F. Reuleaux mechanism that has a scientific value, all details of the model of the mechanism will be repaired, obvious and implicit defects are eliminated. It is planned that restoration process of the model of the mechanism will be made conducted gradually and will last for 3 months (Fig. 4).

The specifics of the task of determining the cost-effectiveness of restoration work of the Franz Reuleaux mechanism is that the historical heritage, as a rule, can not be measured by an economic criteria. It is predominantly a social value. Therefore, well-known recommendations of the evaluation of the effectiveness of investments in these conditions should be modified [5, 6].

Fig. 2 The mechanism S8 (Q-17)

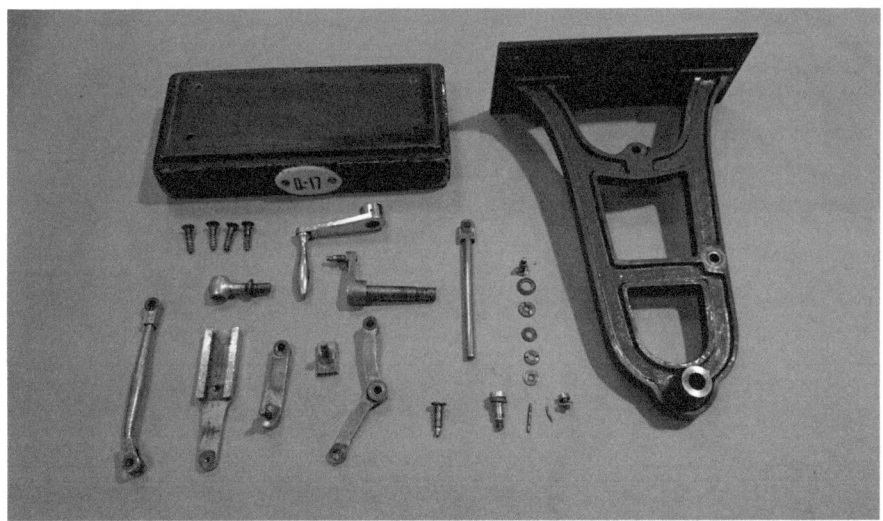

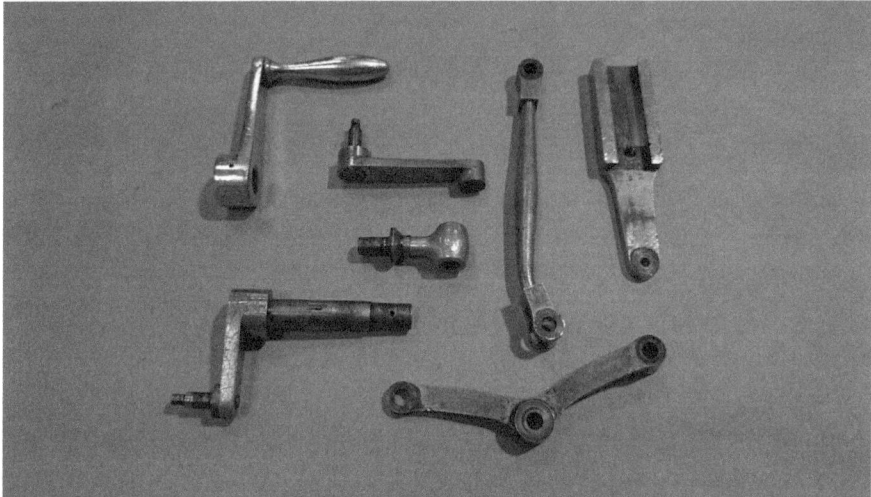

Fig. 3 Components of the mechanism S8 (Q-17)

After determining the value of each object (the model of the mechanism) in points and knowing the costs of restoration of a model of a F. Reuleaux mechanism, we can determine the cost per unit value of any object.

Then the values found are ranked in the order of decreasing of the costs per unit values and a rational set of the objects that should be restored in the first place is formed.

The estimate documentation is approved before the start of restoration work. The team performing the work with the models of the mechanisms has the opportunity

Fig. 4 Three-dimensional computer model of the mechanism of the F. Reuleux collection

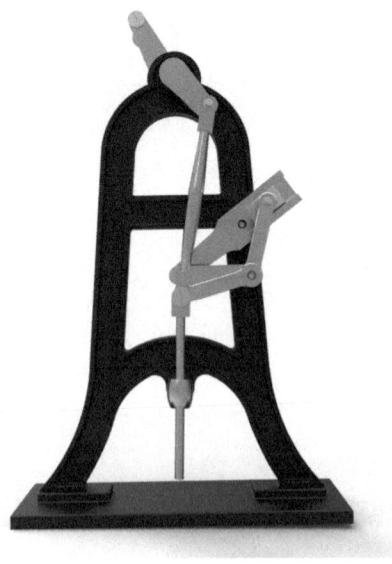

to make the additions and changes to the estimate documentation by submitting proposals to reduce the cost, improve the quality and reduce the duration of restoration process.

The cost of repair and restoration process of the models of the F. Reuleaux mechanisms includes the costs associated with the development of scientific and project documentation, scientific management, organization of work as well as the need to provide the expenses on paying the restorers, artists-restorers, service staff of the complicated technical equipment (3-D printing systems, etc.), modelers, programmers, engineers and designers. It is necessary to provide the expenses for the special equipment in a context that it should be provided with the necessary materials (cartridges of the basic and supporting materials for the technology of fast prototyping, etc.) [7, 8].

The current system of the pricing and the estimated valuation includes the costing standards and other regulatory documents required to determine the estimated cost of the project for the restoration of the mechanisms. Under the estimated cost we consider a set of resources, including intellectual, material, technical and manpower needed to carry out the repair and restoration work.

After all the stages of the restoration process the model of the F. Reuleaux mechanism Q-17 will be preserved and its prototypes, built using the rapid prototyping technology and implemented on a three-dimensional printing system Objet Eden 250, will be used in the learning process of students at BMSTU.

2 Conclusion

The problem of the restoration of the objects of the historical heritage has not settled itself. It requires the further development. One of the aspects of such problem is definition of sequence of the restoration process of the collection. Present resources—and first of all financial—are insufficient for simultaneous restoration of all models from the collection. That is why it is necessary to solve the problem of the sequence of the restoration process of the models of the mechanisms from F. Reuleaux collection.

Restoration process of a unique technical mechanism is conducted in order to preserve the value of the exhibit and to ensure its safety. During the project the following types of work are done: a visual study of the mechanism, an analysis of its performance, analysis of the possibility of its dismantling, removal, restoration, reconstruction, repair, adaptation, computer three-dimensional modeling, prototyping aimed to create new objects (prototype models of the mechanism), debugging and full-scale test. As a result of the restoration process, students at Bauman University will be able to do laboratory tests and experiments during the course of "Theory of machines and mechanisms", using the prototypes of the F. Reuleaux mechanisms.

References

1. Yakhont, O.V.: Problems of preservation, restoration and attribution of works of art. Featured Articles, p. 463
2. The Ministry of Culture of the Russian Federation, The arch of restoration rules. Moscow (2011)
3. Brekalov, V.G., Terehova, N.U., Safin D.U.: Application of technology of three-dimensional prototyping in educational process, Magazine Design and Technologies, № 29(71), pp. 118–123. Moscow (2012)
4. Brekalov, V.G., Terehova, N.U.: Application of technology of prototyping at creation of physical models from polymeric materials. All materials. Encyclopedic Reference, № 4. C.6–9, p. 118–123
5. Bobrov, Y.G.: The Theory of Restoration of Monuments of Art: Laws and Contradictions, p. 344. p. h. Edsmit (2004)
6. Sokolovsky, D.V.: Algorithms form a rational set of the restored historical heritage. Electronic journal Proceedings of the MAI. № 3, (2000)
7. Brekalov, V.G., Terehova, N.U., Klenin, A.I.: The decision of problems of forecasting and strategic planning of activity of higher educational institutions. Eur. Soc Sci. J. № 4. T.2, 31–35 (2014)
8. Brekalov, V.G., Terehova, N.U.: Application of technology of prototyping at creation of physical models from polymeric materials. Polym. Ser. D, № 4 (2015)

Application of Rapid Prototyping Technology for Modeling the Mechanisms of F. Reuleaux Collection

M. Vlasov and M. Samoylova

Abstract This article examines the possibility of the application of rapid prototyping technology for modeling the mechanisms of F. Reuleaux collection stored in Bauman Moscow State Technical University (BMSTU). Nowadays some models from the collection are successfully being used in the students' learning process in order to explain the structure, the kinematics and the dynamics of the mechanisms. However, other models from the collection over time have lost their efficiency so at the moment they are in need of the restoration. A study of F. Reuleaux model S8 (Q-4) was conducted by the analyzing of all its constituent units as well as computer 3D-modeling which allowed us to obtain a prototype of the mechanism with the use of the three-dimensional printing systems: Eden 250 and ZPrinter 650.

Keywords F. Reuleaux · Rapid prototyping · Kinematic schemes · Theory of mechanisms and machines · Restoration · 3D-modeling · Three-dimensional printing

1 Introduction

The department of "theory of mechanisms and machines" is one of the oldest not only in the Bauman Moscow State Technical University but also among the leading technical universities in Russia and abroad. Created by the outstanding scientists, its history goes back over 140 years. The department has a unique cabinet of mechanisms, founded by I.P. Balashev in 1845, which include more than 500 models that are widely used in the educational process. The collection includes more than fifty models of the mechanisms of Franz Reuleaux collection [1].

M. Vlasov (✉) · M. Samoylova
The Bauman Moscow State Technical University, Moscow, Russia
e-mail: maximvlasov@inbox.ru

M. Samoylova
e-mail: cm2003@list.ru

Fig. 1 The models of the mechanisms from the collection of Bauman Moscow State Technical University

Application of Rapid Prototyping Technology ...

Theory of mechanisms and machines is the scientific discipline that studies the structure, kinematics and dynamics of mechanisms in connection with their analysis and synthesis. The objective of the structural analysis is the problem of determining the parameters of the structure of the—the number of units and the type of kinematic pairs, etc. The problem of structural synthesis is the synthesis of the mechanism with the desired properties.

The use of the mechanisms from F. Reuleaux collection in the learning process enables students to understand the structural synthesis and analysis of the mechanisms. Some of the mechanisms are presented at Fig. 1.

Nevertheless some models' operating property is gradually deteriorating due to the wear of parts, that is why they are in need of restoration that should be carried out based on the rules and regulations in the field of restoration of unique historical technical objects [2–4].

One of the models of the mechanisms from the Franz Reuleaux collection of Bauman University—model S8 (Q-4)—is shown in Fig. 2. Model S8 is based on dual slider mechanism, in which two prismatic guides are not at right angles.

This mechanism was used during the laboratory works of the discipline 'Theory of mechanisms and machines' by the students of the Bauman University, but currently it is in the process of restoration.

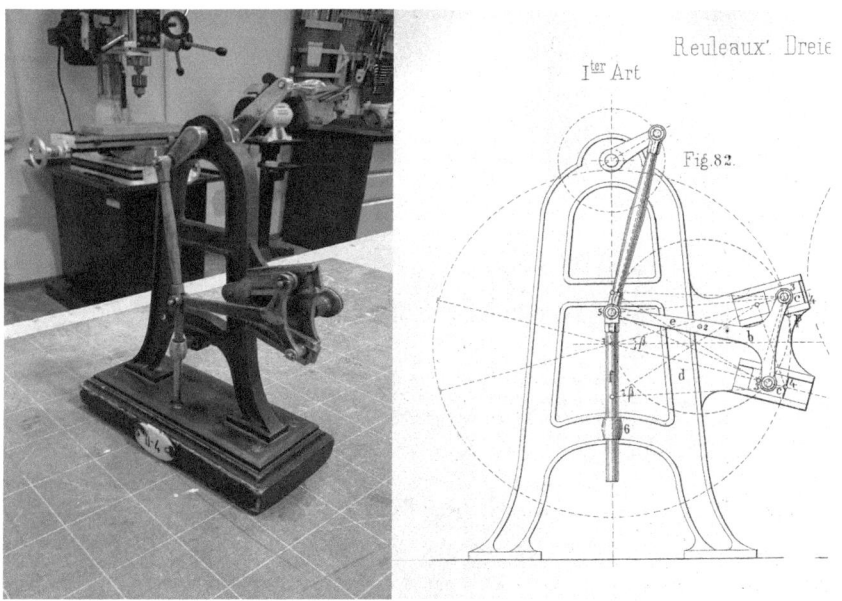

Fig. 2 Model S8 of the mechanisms from the Franz Reuleaux collection (Q-4)

2 3D-Modelling

During the first step of the restoration process we examined the model S8 (Q-4) by analyzing all its constituent units (visual analysis, etc.). Figure 3 shows a model of this mechanism in the form prepared for the subsequent rapid prototyping. This mechanism consists of four significant parts: the two sliders, one link of the drawbar, which traces the curve, and the permanent link, which includes two guides.

During the study it has been found out that it is possible to create a new model of the mechanism S8 (Q-4) with the use of rapid prototyping technology of three-dimensional printing systems Eden 250 and Zprinter 650.

The rapid prototyping technology used to create, as a rule, working models to demonstrate the principle of their action, can be realized in the following ways: stereolithography, curing on a solid basis, coating of thermoplastics, spraying of thermoplastics, or laser sintering of powders [4, 5].

During the first stage of prototyping we constructed a computer three-dimensional model of the mechanism S8 (Q-4), which allowed us to analyze the structure of the virtual model and look for the possible mistakes as well as make the necessary adjustments and changes to the model before the next step of the prototyping process. Moreover, the three-dimensional virtual allowed us to pick up the necessary materials and technologies to create a prototype.

An example of the three-dimensional computer model of the mechanism F. Reuleaux is shown in Fig. 4.

Fig. 3 Components of the mechanism S8 (Q-4)

Fig. 4 Three-dimensional computer model of the mechanism of the F. Reuleux collection

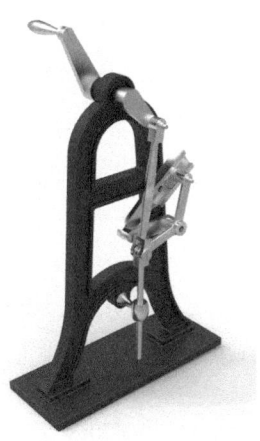

3 3D-Printing

During the next stage of prototyping of the model of F. Reuleaux mechanism S8 (Q-4) we chosen the method of stereolithography realized with the help of three-dimensional printing system EDEN-250, shown in Fig. 5, and the method of the color printing technology using a 3-D printing system Zprinter 650 (Fig. 6). Stereolithography (STL—stereolithography), is one of the methods of rapid

Fig. 5 3D printing system EDEN 250 used for prototyping of the model S8 (Q-4)

Fig. 6 3D printing system Zprinter 650 used for prototyping of the model S8 (Q-4)

prototyping technology that differs from the other 3D-printing methods since it uses the photopolymers in a liquid state as the 'building blocks' instead of the powders [6].

The main advantage of the stereolithography is the high accuracy of printing. This technology allows the printer EDEN-250 to apply layers of a thickness of 15 microns, which is much thinner than the thickness of a human hair. The speed of printing is relatively high, taking into the account the high resolution of such devices: it takes a few hours to construct the model of the Franz Reuleaux mechanism, but in the end depends on the size of the model and the number of laser heads used by the device simultaneously [6].

The working technology of Zprinter 650 is called ColorJet Printing (CJP) which applies model material and the glue in the consecutive layers. CJP belongs to the one of the industrial 3D-printing methods that allows us to obtain full color models of the mechanisms or their parts that can be provide the additional help for the students in the learning process during a demonstration of mechanisms' structure and the principle of their action. 3D-printer Zprinter 650 also provides high accuracy of the models. The smallest element produced by this system has a size of 0.1 mm.

4 Conclusion

Modern rapid prototyping technologies that are used to produce prototypes of the models of the mechanisms will extend the 'life' of historical collections, which include the models of F. Reuleaux mechanisms stored in Bauman Moscow State Technical University. The main advantage of using three-dimension printing systems is the speed of getting a prototype as well as the accuracy of the produced models. Prototypes obtained by three-dimensional printing systems such as Eden 250 and Zprinter 650, will be a unique object for research and study for the students as well as an additional teaching tool for explaining the basics of the theory of mechanisms and machines.

References

1. Golovin, A., Tarabarin V.: Russian Models from the Mechanisms Collection of Bauman University, p. 246. Springer Science+Business Media B.V. (2008)
2. Bobrov, Y.G.: The theory of restoration of monuments of art: laws and contradictions, p. 344. p. h. Edsmit (2004)
3. Sokolovsky, D.V.: Algorithms form a rational set of the restored historical heritage. Electronic journal Proceedings of the MAI. 3, (2000)
4. Yakhont, O.V.: Problems of preservation, restoration and attribution of works of art. Featured Articles, p. 463
5. Brekalov, V.G., Terehova, N.U.: Application of technology of prototyping at creation of physical models from polymeric materials. All materials. Encyclopedic Reference 2015, 4. C.6–9
6. Brekalov, V.G., Terehova, N.U., Safin, D.U.: Application of technology of three-dimensional prototyping in educational process, Magazine Design and Technologies, p. 118–123, 29(71). Moscow (2012)

Protoepistemology of Mechanical Engineering in Cassiodorus' *Variae* or Mission Impossible at Theoderic's Court

N. Ambrosetti

Abstract Starting from the early 6th century CE, the Roman patrician Cassiodorus writes a conspicuous corpus of letters, entitled Variae, on behalf of his sovereign, the Ostrogoth Theoderic, and of his successors. These official letters, intended to show the king's positive attitude towards other members of the court or to foster political relationships with foreign monarchs, offer the well-educated author the opportunity to address high-profile cultural issues. In one letter of the collection, he sings the praises of mechanical engineering, described as one of the most inspiring and powerful arts, competing none less than with nature.

Keywords Mechanics · Engineering · Cassiodorus · Theoderic · Archimedes · Letters · Ostrogoth · Liberal arts · Mechanical arts · Boëthius · Clock · Machine

1 Introduction

An actual notion of epistemology dates back to the modern or even contemporary era. Nonetheless, maybe from an ingenuous standpoint, even philosophers and scholars of ancient times speculated about the nature and the worth of scientific and technological knowledge. In this paper, the 6th century only source on the topic is analyzed, in order to identify interesting traces of such an approach as for mechanical engineering.

N. Ambrosetti (✉)
Università Degli Studi Di Milano, Milan, Italy
e-mail: ambrosetti@di.unimi.it

2 Historical Context

After the fall of the Western Roman Empire, dating 476 CE, Italy was ruled by the Ostrogoths, the Eastern part of the horde of Goths, one of the most important German tribes. They had settled in Pannonia (now Hungary), Balkans, and conquered Italy, where they set up Ravenna as their capital.

From 493 until his death, in 526, Theoderic the Great was the king of the Ostrogothic Kingdom. Born in Pannonia in 454, he had been sent by his father Theodemir to Constantinople as a royal hostage, i.e. as a living guarantee for the respect of political agreements between Ostrogoths and Byzantines. In the capital of the Eastern Roman Empire, he received a refined education.

In 473, he succeeded his father and got the titles of *patricius* (nobleman), *magister militum* (master of the soldiers), and Roman *consul*, appointed by the emperor Zeno, who, 15 years later, sent Theoderic to depose another German king, Odovacer; the emperor had equally appointed Odovacer patrician and also King of Italy, trying to push the two to fight and destroy one another. Nonetheless, Theoderic defeated his enemy (he is said to have literally beheaded Odovacer), and with his people, ranging from 100,000 to 200,000 individuals, founded a kingdom on the ruins of the Western Empire [1, 2].

Ravenna was the ideal capital for the new-born kingdom, as a sign of continuity with the fallen Roman Empire, and for the presence of the city's harbor in Classe, and of important trade routes to the East.

Theoderic commissioned the construction of many impressive buildings both in Ravenna, such as the (now lost) royal palace, churches, a cathedral and a baptistery; and also outside the city walls, like his own Mausoleum, one of the few surviving monuments.

So far, Theoderic may appear as a religious king, respectful towards Roman tradition, submissive to the Eastern emperor, loyal to his people, devoted to the prosperity of his reign. But, at a closer look, this first impression of universal harmony disappears, giving way to the consideration of the multifaceted king's policy [3, 4].

First of all, Theoderic and his people were Arian Christians, while the Latins were mainly Orthodox: this difference forced him to build a dedicated church and a baptistery for the Goths, in order to ensure everyone a proper place of worship.

Then, Theoderic applied a double system within his kingdom: while Roman citizens remained subject to their legal tradition, the Ostrogoths insisted on preserving their own laws and customs. Such a separation was not only tolerated, but also promoted by the king, with the prohibition of mixed marriages.

Again, despite the apparently good relationships with the Empire, the king was constantly looking for new alliances with other barbaric peoples, such as the Burgundians, or the Franks, in order to strengthen his power in the Western lands (Fig. 1).

In addition, from the cultural perspective, the two peoples were very dissimilar; the Latins, despite their long history of imperialism, appeared principally devoted to

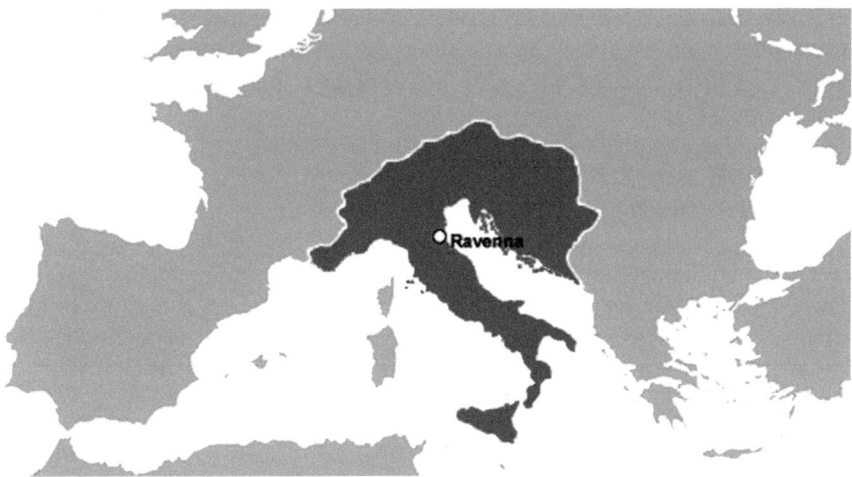

Fig. 1 The Ostrogothic kingdom

liberal arts, while the Goths showed talent mainly for warfare. Again Theoderic decided to keep the two peoples separated from one another; the Romans could devote themselves to the kingdom administration, to bureaucracy, and to the preservation of the ancient philosophical and scientific culture, under close supervision of the king, while the more reliable Goths would have held the exclusive control of military power [5].

The wonderful, albeit altered, mosaic preserved in the church of Sant'Apollinare Nuovo, in Ravenna, shows impeccably the admiration of the king for the Latin cultural environment. As shown in Fig. 2, the building has both a Latin name (the word Palatium had been used for the first time to refer to Augustus' Palace on the Palatine hill in Rome) and a typical Latin architecture, a series of columns and arcs.

Fig. 2 Theoderic's palatium (mosaic in Sant'Apollinare Nuovo, Ravenna)

Under each arc, a figure of a Theoderic's courtier appeared, while in the central arc, we can suppose that the king was sitting in his throne. The absence of figures is now due to the Byzantines, who, after the fall of the Ostrogoths, replaced them with curtains, as a kind of *damnatio memoriae*.

Among those courtiers, two prominent Roman statesmen, whose lives are strictly linked, could have drawn our attention: Anicius Manlius Severinus Boëthius (c. 475–524) and Flavius Magnus Aurelius Cassiodorus Senator (c. 490–c. 585).

Boëthius, belonging to the Anicii family, was part of the highest senatorial nobility, and linked to the political party that had strong influence both on Theoderic's, and on the Eastern Roman Empire's court. This group looked at the king as at a point of convergence, who could in fact achieve a balanced policy both inside and outside the kingdom, mediating among the Latins and the Goths (Figs. 3, 4 and 5).

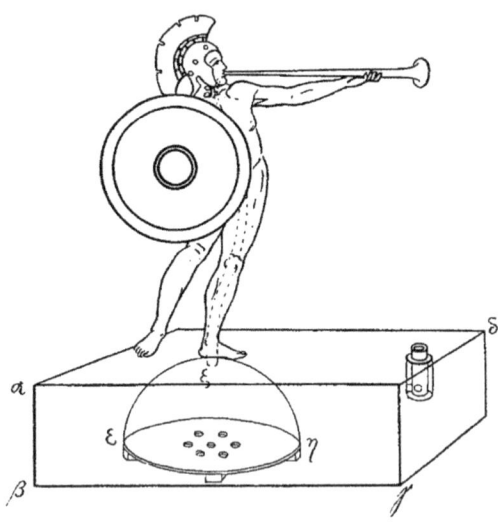

Fig. 3 Heron's device #II, 11 (according to Schmidt's numbering)

Fig. 4 Philon's device #58 (according to Carra de Vaux' numbering)

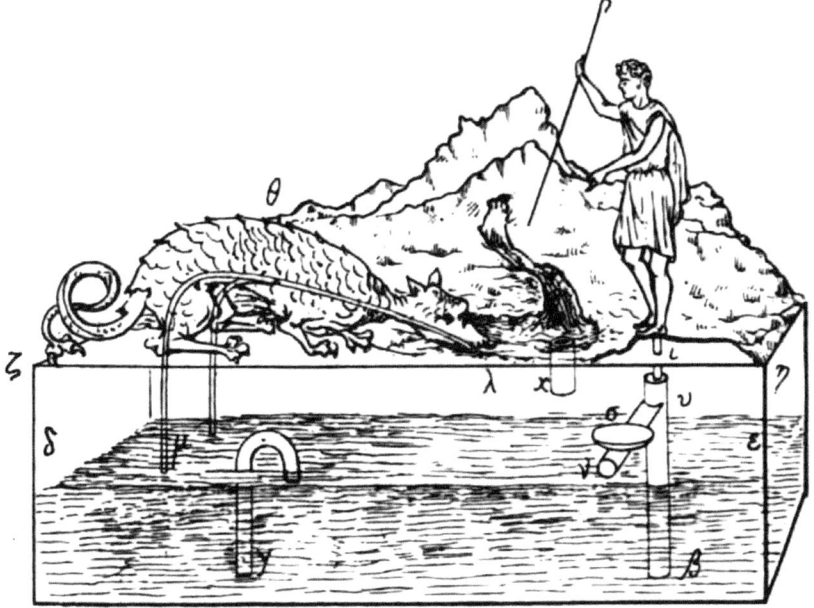

Fig. 5 Philon's device #59 (according to Carra de Vaux' numbering)

In the same years, the younger Cassiodorus was following Boëthius' footsteps in protecting and studying classical works, mainly focused upon theology. At the court, however, he worked as the king's secretary, a real ghost writer of the royal correspondence, that he later collected in his work *Variae*, our main source [6–8].

As Boëthius was giving a strong contribution to Theoderic's policy, he was repeatedly asked by the king to provide the court with his knowledge on the liberal arts of the *Quadrivium* (arithmetic, geometry, music, and astronomy). Therefore, he translated the major works of ancient Greek mathematicians, philosophers, and scientists.

This is the reason why he was called in 496 to choose a lyre to be sent as a gift to the Frankish King Clovis (*Variae*, II, 40); in 509, again, requested by Theoderic, he presided over the construction of an astronomic clock for Gundobald, king of the Burgundians (*Variae*, I, 45) [9–11]; as an expert in arithmetic, he also used to oversee the accounting of salaries of the royal employees (*Variae*, I, 10).

Boëthius' cultural task consisted mainly in preserving the great achievements of Greek culture in Latin language; he was strong of his prestige, as one of the best counselors of Theoderic: he tried therefore to influence the king's policy, fostering the links with the Latin tradition. Unfortunately, in 522, a conflict of political, and possibly religious, nature dropped Boëthius out of the king's favor: he was charged with treason, imprisoned, and executed in 526.

3 Mechanics and Machines in Variae

Before deepening the definition of mechanics' nature in a specific letter, it is worthwhile to point out that references to the theme are present in the whole collection, though with different meaning nuances.

3.1 Linguistic Analysis

In Cassiodorus' letters various words are etymologically related to *machina*: he speaks about objects such as *machina* (3 occurrences), *machinamentum* (1), *mechanisma* (1), meaning machines. The 3 synonyms are all present at I, 45, while *machina* occurs also in II, 40 (*machina mundi*: the universe, whose melody can be reproduced by the lyre built by Boëthius) and at V, 42 (a revolving wheel, used during gladiators' fight in the Colosseum).

Cassiodorus uses the word *mechanicus* as a title for Archimedes (I, 45), or in the sense of mechanician (I, 45; III, 52; III, 53); the adjective *mechanicus* also occurs in the expression *res mechanicae*, meaning applied mechanics (III, 52). On the contrary, the word *machinator* is used to denote false accusers in a trial (1 occurrence at IV, 41). The term used to denote technicians able to put into operation a machine (namely a clock) is *magistri rerum* (masters), or *dispositores* (installers).

In the last books, Cassiodorus insists with the same negative nuance, related to the word *machinatio* (machination). It occurs 6 times (III, 15; V, 39; VI, 13; VIII, 16; IX, 15 and 18) in letters and public documents, all dating after the fall of Boëthius and ranging from 523 to 533, in a curious coincidence.

We also considered the occurrences of the word denoting the device built by Boëthius: *horologium* (etymologically, a contrivance that allows to "read" what time it is). As a matter of fact, this term appears in the letters only at I, 45 and I, 46 (an accompanying note for the gifts), but it is also used in Vitruvius' De architectura [12] to mention both sundials (IX, 1), and clepsydras (IX, 8).

3.2 Variae, I, 45

The most intriguing reference to our theme is at Variae I, 45, which is the only source about the topic in the 6th century. The letter dates back to 507, when Boëthius was at the top of his prestige with Theoderic. A thorough summary of the content is here given with some minor notes, in order to allow a complete understanding of the content details.

The king starts his letter by referring to strategic reasons, which force him to unwillingly accept (spernenda non sunt) presumptuous requests of small gifts (res parvae) from close (vicinis) monarchs. Through these minor concessions to playful presents, he hopes to achieve serious goals, inaccessible via war (arma), or wealth (divitiae). Theoderic is then, on the one hand, emphasizing the political importance of his request, while, on the other hand, he is describing the foreign king's wish as not so hard to please.

The author of the request is the "dominus" (interestingly not referred to as a king) of Burgundy, who has asked in fact a triple gift: a water clock, a sundial, and skilled workers to set them up. Such a demanding pretense can be explained only by the kinship between the two kings: Theoderic's daughter Areagni was married with Sigmund, son of the requester. The king underlines the superiority of the Goths on other barbarians: such devices are common (cottidianum) for his people, but they will however appear a miracle to the Burgundians, who have only heard about them from the accounts of their ambassadors. Possibly feeling some annoyance for the request, Theoderic wanted to underline the distance of his reign from the others' in front of Boëthius, who was a keeper of the Greek and Roman culture.

In the 3rd paragraph, the king acknowledges all Boëthius' qualities, starting from his competence in disciplines, that are normally (vulgariter) practiced by uneducated people (nescientes). Theoderic continues describing the huge effort made by Boëthius: making Greek scientific and philosophical studies (Graecorum dogmata) part of the Roman culture (doctrinam), by translations. In the 4th paragraph, the translated works are listed in couples: Pythagoras' music, and Ptolemy's astronomy; Nichomachus' arithmetic, and Euclid's geometry; Plato's theology and Aristotle's logic. This choice is very noteworthy, because it appears an actual curriculum studiorum, composed by the four disciplines of the quadrivium (term coined by Boëthius) and by philosophy. The reference to the translation of Archimedes' mechanics is the last, separated from the other ones (mechanicum etiam Archimedem Latialem Siculis reddidisti), as a last but not least author. The recognition of excellence of the work by Boëthius is extended to his elegant style, and to his appropriate vocabulary: if only the Greek authors could choose between the original and the Latin version, they would pick out the second one.

In the 5th paragraph, the king begins his description of the nature of mechanics, which requires the knowledge of the liberal arts of the quadrivium (quadrifarias mathesis ianuas). At the king's eyes, the genius of Boethius was able to take advantage of the reading of the aforementioned Greek treatises to build his skills in mechanics, which is based upon the mysterious laws of nature (naturae penetralibus). Mechanics is therefore strictly devoted to wonder (miraculum), according to Theoderic: it can completely control the sequence of the events, in order to inspire awe and, at the same time, it also prevents from believing in what is under the eyes (cum ostendet et oculis visionem). Theoderic then links the usage of mechanics with three elements: water, fire, and air. The first citation alludes to water-lifting (facit aquas ex imo surgentes praecipites cadere, cfr. Vitruvius X, 12: machina quae in altitudinem aquam educit); then to its fall on fire thanks to its weight (ignem ponderibus currere); and finally to air flow in the organ pipes (organa ... insonare, et peregrinis flatibus calamos complet).

As Theoderic adds in the 6th paragraph, by means of mechanics, which takes advantage of physical laws, useful appliances have become astonishingly possible: the defenses of a city already about to collapse, immediately bounce back to victory with the help of machines (machinamentorum auxiliis); waterlogged structures (madentes fabricae) can be drained though still in the sea; even the hardest materials can be broken thanks to an ingenious device (ingeniosa dispositione). In addition, some entertaining usages are quoted. Metal objects can emit deep sounds apparently by themselves (metalla mugiunt): a bronze Diomedes can play a note (bucinat) at the trumpet, a bronze snake can hiss (aeneus anguis insibilat), models of birds (aves simulatae) can chirp, and objects without a voice of their own (vocem propriam nesciunt habere) can give out a sweet melody (dulcedinem cantilenae). As we will later discuss, such a distinction between useful and playful devices has not been accidentally made.

In the 7th paragraph, Theoderic feels the need to return on the essence of mechanics, as he has not adequately described the nature of this art in the previous paragraphs: mechanics can imitate the sky (caelum imitari fas est), as it allowed to build a small model of the celestial bodies' reciprocal movements. The author explicitly refers to Archimedes' armillary sphere, which included the sun, the moon, and the zodiac (alterum zodiacum circulum), in movement: humans cannot perceive celestial bodies' movements, as they appear still to human eyes, but thanks to this "portable heaven", "compendium of the universe", "mirror of nature" (caelum gestabile, compendium rerum, speculum naturae), a deeper comprehension of astronomic phenomena is at hand. Despite this explicit mention of Archimedes' sphere, as we will later show, we have no evidence of a direct knowledge about the topic, but the quotation is possibly inspired by a secondary source.

After this long description of the extraordinary achievements due to mechanical knowledge, in the 8th paragraph, Theoderic finally asks his courtier to build the clocks at public expenses, detailing the requirements: the first one must be a sundial, where a thin gnomon casts its shadow. The author describes his wonder at comparing the huge size of the sun, which moves along its orbit, to the small unmoving circle of the sundial, which, however, perfectly reproduces the movement of the sun. Again Theoderic emphasizes the contrasts, inherent in the realization of such a device. Besides the abovementioned macroscopic differences, in the 9th paragraph, the sundial working principle is described as an actual paradox; he suggests to consider that, while daytime is due to the light (horarum de lumine venientium singulare miraculum), information about the hour is displayed by a shadow ([horas] umbra demonstrat); in addition, in a sundial the fast movement (indefecta rotatio) of the sun along the horizon is reproduced by a metal piece fixed in a stable place (metalla situ perpetuo continentur).

As for the second clock, in the 10th paragraph, Theoderic underlines that its operation must not depend on the sunlight, and that it must work during the night, too. The king is referring to a clepsydra, as he describes its working as changing the skies into streams of water (rationem caeli ad aquarum potius fluenta convertit). The power to perform such a conversion, according to Theoderic, demonstrates the distinctive quintessence of mechanics: while the other scientific disciplines only aim at understanding (nosse) the power of nature (naturae potentiam), mechanics (mechanisma) dares to imitate nature by implementing the opposite, or even, if one may say so, it also tries to overcome nature under some respects, as we will later discuss.

The author adds some examples related to flight, in order to prove such a challenge to nature: Daedalus, who was able to build wax wings, and to fly (Daedalum volare); and an iron statue of Cupid, floating in the air in the temple of Artemis in Ephesus (Cupidinem in Dianae templo sine aliqua illigatione pendere). The two examples are taken by Theoderic from Greek mythology, in order to celebrate the extraordinary power of the discipline in ancient times; today, the king adds, mechanics can transform the intrinsic characteristics of objects in their opposites: it can make mute things sing, inanimate live, immobile move (muta cantare, inanimata vivere, immobilia moveri).

If this is the power of mechanics, continues Theoderic in the 11th paragraph, a scholar who practices it (a mechanicus), is very different from the other scientists. He must be considered a nature comrade (socius naturae), because he can not only unveil the secrets of nature (occulta reserans), but, at will, he also uses natural laws in order to create a wonderful, but artificial and trustworthy imitation of reality (quod compositum non ambigitur, veritas aestimetur). Boethius is then required by the king to send the clocks as soon as possible, in order to become famous in a part of the world his name would not be able to reach otherwise.

The final paragraph of the letter emphasizes again the political opportunity given by the Burgundian king's request: his people will wonder at Boethius' clocks, and, when their wonder is over, they will not dare to consider themselves equal to the Goths, because they know that renown scholars, who designed such important devices, are only in Theoderic's reign.

4 Protoepistemology of Mechanics

In this letter, Theoderic takes the opportunity given by a common political event, to tackle some interesting protoepistemological topics related to mechanics, and to introduce a definition of "mechanicus", the ancient instance of mechanical engineer.

4.1 *Liberal and Mechanical Arts in Variae I, 45*

In the first part of the letter, the author considers both liberal, and technical arts: at the king's eyes, Boëthius appears remarkable, because he is skilled not only in the liberal arts, but also in the mechanical ones, i.e. those involving a physical labor. Such an opposition dates back at least to Cicero's *De oratore*, where the intellectual arts are defined as the only ones suitable for a free man, while slaves practiced the mechanical ones. Cicero's taxonomy of knowledge has been completely transformed: far from a dualistic perspective, the author proposes a unitary vision of knowledge, and assigns a prestigious task to mechanics. Theoderic defines liberal arts as a kind of preliminary step (*quadrifarias ianuas*) required to be introduced to a more difficult area of knowledge (*mathesis*), dealing with the mysterious laws of nature (*penetralia*) and devoted to show wonders (*miracula monstrare*). This idea of mechanics is more based on the Greek culture than on the Latin one, and appears recalling both the introduction of Pseudo-Aristotle's *Mechanics,* and the treatises about pneumatics and automata of the school of Alexandria.

4.2 The Role of Wonder

The introduction of Pseudo-Aristotle's work opens with the verb to wonder (θαυμάζειν) referring to the first step of investigation: anyone who wants to understand a phenomenon, begins by wondering. Phenomena however are of a double kind: natural (κατὰ φύσιν), due to unknown causes, and artificial (παρὰ φύσιν), due to technical skills and devoted to usefulness (τὸ χρήσιμον), the same goal (*utilitas*) included in Vitruvius' canon for any building.

In the letter, Theoderic portrays Boëthius as a scholar who has already passed the stage of wondering about natural phenomena, due to his complete scientific knowledge; like the ancient Alexandrian engineers, quoted through their works, he can use his technical skills, in order to make other people wonder, while admiring his contrivances. In addition, Theoderic explains how this goal is achieved: wonder comes, first of all, from the consideration that, in a mechanical device, a given cause produces an opposite effect (e.g., a tiny object, such as the armillary sphere, can display the universe; a shadow can display the hours, which depend on light). Showing the invisible, such as the movement of the stars; and hiding causes which produce the effect (organ that emit notes), or even making possible the impossible (the flight of Daedalus or the floating of Cupid) complete the list of ways to achieve wonder. Theoderic does not hide that, in this specific case, such a wonder will be used for political goals: emphasizing the power of the Goths at the Burgundians' eyes.

4.3 Usefulness and Playfulness

In the long list of the quoted mechanical devices, only a small number can be considered useful *stricto* sensu, while most of them are identical to the automata described in Heron's [13] or Philon's [14] treatises and built mostly for entertainment.

As previously referred, in the last book of Vitruvius' De architectura, some interesting examples of usage of the suction and discharge pump, allegedly invented by Ctesibius, are included: machines used in civil and military engineering, and also the hydraulic organ. Vitruvius' list fairly corresponds to the one in the 6th paragraph of Variae I, 45: on the one hand, the useful devices, such as military defenses for a besieged city, structures (possibly ships) to drain, materials to break; on the other hand, the playful ones, metal objects emitting deep sounds, statues replicating instruments or reproducing animal sounds, or even an object emitting a melody, probably an organ.

Even though Vitruvius has never been explicitly quoted by Theoderic it is curious to underline that the De architectura X, 7 includes a reference to a playful usage of water pressure by Ctesibius (merularum aquae motu voces atque angobatae bibentiaque et eadem moventia sigilla, ceteraque quae delectationibus oculorum et aurium usu sensus eblandiantur—imitate the voices of singing birds, and the engibita, which move figures that seem to drink, and perform other actions pleasing to the senses of sight and hearing). In the last paragraph of his treatise, Vitruvius explains that he has selected the most pleasing and necessary inventions by Ctesibius, and he invites the curious to read directly the writings of Ctesibius to find many other examples.

As we can see, while Vitruvius' goal was utilitas, for him it is not necessary to expose the playful part of Ctesibius' treatise, for Theoderic such an opposition makes only apparently sense. As the king's aim is to impress Gundobald with the superiority of his courtier's skills, utilitas and voluptas are here not in conflict, rather contribute to the same political result.

4.4 Mission: Impossible

The end of the letter is also very interesting, because another difference between sciences and mechanics is put into evidence: their relationship with nature. While scientists are focused upon investigating natural phenomena, in order to understand and explain their causes, mechanics goes further. It masters natural laws and is able to replicate natural phenomena, in order to achieve unexpected, astonishing, and apparently non-natural effects, such as a statue floating in Diana's temple, or metal objects singing. Mechanics, in a sense, allows to minimize the predictability of an effect from its cause.

5 Boëthius as an Engineer?

Ancient Alexandrian engineers' works are not explicitly cited by Theoderic, while Archimedes' *Mechanics* is in the list of the translations made by Boëthius. Unfortunately, this Latin version was not handed down, no other source is available about its real existence, and consequently about Boëthius' real knowledge about the topic.

A passage of the letter possibly refers to the armillary sphere by Archimedes: carried off from Syracuse by Marcellus after Archimedes' death in 202 BC, owned by Sulpicius Gallus, the planetarium was later described by Cicero in his *Tusculanae Disputationes* I, 63, and in *De re publica* I, 21–22. Even though the works by the greatest Latin orator can be considered as a likely source for the letter, it would have been impossible for Boëthius to build a real and working celestial sphere merely based upon a literary description, totally lacking of the required technical specifications [15].

Archeological finds, such as the Antikythera mechanism or the gearwheel found in Olbia in 2006, testify that Greek technology was advanced enough in the second century BC, more or less eight centuries before Theoderic. More recent evidence is not available in the Western Roman Empire.

Due to this lack of sources, we could wonder whether the construction of the clock was actually feasible in the Gothic age. A first answer is partly enclosed in the letter: maybe Theoderic exaggerates when he defines the clock as an ordinary device for the Ostrogothic people, but it must have been at least a device they were aware of. The request of technicians to put the clock into operation is another indirect confirmation that such masters actually existed.

Lastly, Variae I, 46 gives possibly the final answer to the question: the letter is sent by Theoderic to Gundobald and it is an accompanying note for the clocks. Even though we cannot be sure that it has really been sent, it is difficult to argue that Cassiodorus, who years later collected the letters, could have included in his work the evidence of a king's failure.

6 Conclusions

Variae I, 45 by Cassiodorus is one of the rarest sources about technical knowledge in Western Roman Empire in the 6th century. We must also admit that the *Variae* are mainly a rhetorical work, written to impress the receivers, and therefore composed taking into account as many diplomatic and political issues as possible. In simpler words, the letters must be considered a secondary source. In addition, Boëthius' works about the liberal arts can testify their author's scientific skills, but not his actual technical expertise.

Nonetheless, at least as for our specific passage, we can consider Cassiodorus (and Boëthius, at turn) an eye witness of a crucial turning point: the end of the ancient world, the transition to another, less refined culture [16]. Boëthius even felt a need for preserving Greek and Latin knowledge at any cost: while being a philosopher, he probably even became a "mechanicus", a protoepistemological instance of a mechanical engineer.

References

1. Ward-Perkins, B.: The Fall of Rome and the End of Civilization. Oxford University Press (2005)
2. Wozniak, F.E.: East Rome, Ravenna and Western Illyricum: 454–536 A.D. Historia: Zeitschrift für Alte Geschichte **30**, 351–382 (1981)
3. Vitiello, M.: Il principe, il filosofo, il guerriero: lineamenti di pensiero politico nell'Italia Ostrogota, Franz Steiner Verlag (2006)
4. Sirago, V.A.: Gli Ostrogoti in Gallia secondo le Variae di Cassiodoro. Revue des Études Anciennes **89**, 63–77 (1987)
5. Punzi, G.A.: L'Italia del VI secolo nelle Variae di Cassiodoro (Series: Piccola biblioteca di cultura). Vecchioni (1927)
6. Mommsen, T. (ed.): Cassiodori Senatoris Variae (Series: Monumenta Germaniae Historica). Weidmann (1894)
7. Aricò, G.: Cassiodoro e la cultura Latina. In: Atti della Settimana di Studi su Flavio Magno Aurelio Cassiodoro (Cosenza-Squillace 19–24 settembre 1983), pp. 154–178 (1986)
8. O'Donnell, J.J.: Cassiodorus. University of California Press (1979)
9. Saitta, B.: I Burgundi: (413–534). Muglia (1977)
10. Pizzani, U.: Boezio consulente tecnico al servizio dei re barbarici. Romanobarbarica: contributi allo studio dei rapporti culturali tra mondo romano e mondo barbarico **3**, 189–242 (1978)
11. Moorhead, J.: Boëthius and Romans in Ostrogothic Service. Historia: Zeitschrift für Alte Geschichte **27**, 1–21 (1978)
12. Gros, P. (ed.): Vitruvio. De Architectura. Einaudi (1997)
13. Schmidt, W. (ed.): Heronis Alexandrini Opera Quae Supersunt Omnia, vol. 1, Pneumatica et automata. Teubner (1899)
14. Carra de Vaux, B.: Le livre des appareils pneumatiques et des machines hydrauliques de Philon de Byzance d'après les versions arabes d'Oxford et de Constantinople. Académie des inscriptions et des belles lettres: notices et extraits des mss. de la Bibliothèque Nationale **38**, 27–235 (1903)
15. McCluskey, S.: Boëthius' astronomy and cosmology. In: A Companion to Boëthius in the Middle Ages, pp. 47–74. Brill (2012)
16. Gillett, A.: The purpose of Cassiodorus' Variae. In: After Rome's Fall: Narrators and Sources of Early Medieval History, pp. 37–50. University of Toronto press (1998)

The Silk Mill "alla Bolognese"

C. Bartolomei and A. Ippolito

Abstract The silk mill of Bologna is an important example of the proto-industrial system of factory production. The present paper examines the distribution of silk mills in Bologna in an attempt to follow its historical development and delineate their material and technological structure. The technology which underlies the functioning of the mills is analyzed while emphasizing the unique nature of these artifacts in the history of machinery. Focus is basically placed on the structure/machine aspect of the mill bringing out certain aspects of the history of technology and material culture.

Keywords Bologna · Silk mills · Hydraulics · Machines · Technology

1 Introduction

The silk mill "*alla Bolognese*" is an important example of the proto-industrial system of factory production. Its technological structure remained unchanged from the XIII until the XX century. It is thus a machine of the longest lifetime in history of machinery. From the very beginning the process of production was completely mechanized. The task of the workers was solely to power thee machines, tie up threads that broke off, take skeins already twisted and place them in their proper baskets. They were provided with mechanical winders and water wheels whose speed was kept constant by a regular flow of water.

It seems somewhat strange that the Renaissance architects had never really analyzed the subject of water power treated with such attention in ancient treatises.

C. Bartolomei
Department of Architecture, University of Bologna, Bologna, Italy
e-mail: cristiana.bartolomei@unibo.it

A. Ippolito (✉)
Department of History, Representation and Restoration of Architecture,
Sapienza University of Roma, Rome, Italy
e-mail: alfonso.ippolito@uniroma1.it

© Springer International Publishing Switzerland 2016
C. López-Cajún and M. Ceccarelli (eds.), *Explorations in the History of Machines and Mechanisms*, History of Mechanism and Machine Science 32, DOI 10.1007/978-3-319-31184-5_4

Vitruvius dedicated his VIII book entirely to hydraulics while the X book of the De Re Edificatoria of Leon Battista Alberti analyzes the subject of water just like his successors Francesco di Giorgio, Leonardo, Taccola, Filarete.

Water powered machines became the subject of serious study only in the XVII century with the outstanding engineers-architects like Vittorio Zonca, Agostino Ramelli, Giambattista Aleotti. Certainly Bologna was a city provided with a great variety of hydraulic systems, which—however—never caught the collective imagination. Bologna is generally associated with the "two towers" almost completely ignoring its industrial and commercial part which developed underground into a network of sluices meant to provide water wheels with a constant flow of water, but which were also visible on the ground where canals and silk mills were constructed.

In point of fact Bologna is not situated on any single river. It lies at the foot of the Apennines and takes advantage of three water streams [7]: Aposa, Savena and Reno. The latter brings water and water power to the north-western parts of the city and served as the main source of water power for the industry of Bologna. The river entered the city though "la Grada", a huge gate in the form of grating, then flowed to the west where it divided into two streams: the Cavaticcio canal directed to the north towards city walls, and the other one—flowing eastwards until it joined Aposa flowing northwards. They both joined in the Moline canal to form a series of cascades marked by pairs of mills on both sides.

The Reno canal used to feed a series of mechanical factories, the most important of them being those involved in the production of silk. Little of the industrial part of the city survived until our day and yet it definitely forms an important part of the heritage not to be ignored.

Spinning and weaving machines were powered by the waters flowing through the canal, which for greater extensiveness of distribution were directed into small underground canals leading through the cellars of houses. The technology developed and applied was so important to the city that it was declared to be a state secret. This explains the scarcity of graphic and written materials concerning the prime textile industry of Bologna. The development of silk production was possible thanks to the introduction of innovative technologies—like the water wheel and the mechanical winder—as well to the intellectual effort of eclectic personalities. Domenico Guglielmini and Luigi Ferdinando Marsili were the protagonists of the cultural and political battles of the XVII and XVIII centuries.

2 Mills in Bologna

The mill [1] (Italian: *mulino*, from the Latin "*molinum*" derived from "*mola*") is an instrument that performs mechanical work taking advantage of power provided by electric energy, wind, water or animal/human force. Initially used for grinding grain, they were later applied in the production of silk. Water powered mills of Bologna [5] were mentioned for the first time around the year 1100 and refer to their construction on the Savena and the Reno rivers. The first mills of Bologna [3] were constructed a little later than those in other parts of the Po Valley. They were involved in the production of bread but due to high costs of construction and maintenance they belonged to aristocrats and ecclesiastic institutions. Only after the battle of Legnano of 1176 the Town Council decided to direct the Savena and the Reno towards Bologna. So, numerous mills were constructed at the cost of the Town Council and then bestowed to individual people. Around the year 1220 private mills on the Reno canal were expropriated to the double advantage for the Council: one, to guarantee to the authorities the profits from letting out the mills, and the other one—to make it impossible for private owners to interfere with the system of providing water. Between the thirteenth and the seventeenth century we witness an almost stable distribution of activity in the city center. Grain grinding mills were concentrated on the Moline canal, north-west of the city and were privatized at the beginning of the fifteenth century. In the southern part of the city the predominant factories were those of woolen mills and dyeing plants, while in the north western zone, watered by the Reno canal, was concentrated the most productive part of the city. Between La Grada, via Lame and the Cavaticcio were situated the majority of silk spinning wheels and all manufactures involved in the production thereof that took advantage of water power, terrain declivity and short distance to the port.

The waters of the Savena seem particularly suited to textile dyeing. Many dyeing plants and activities connected with leather and precious stone working were centered in this zone. The original names of some of the streets have been preserved to this very day: via Dell'oro, via Arienti, via Pellacanti, via Cartolerie, suggesting manufacturing activities followed there (Figs. 1, 2, 3, 4 and 5).

Fig. 1 Silk mill's Heinrich Schickhardt (XVI century)

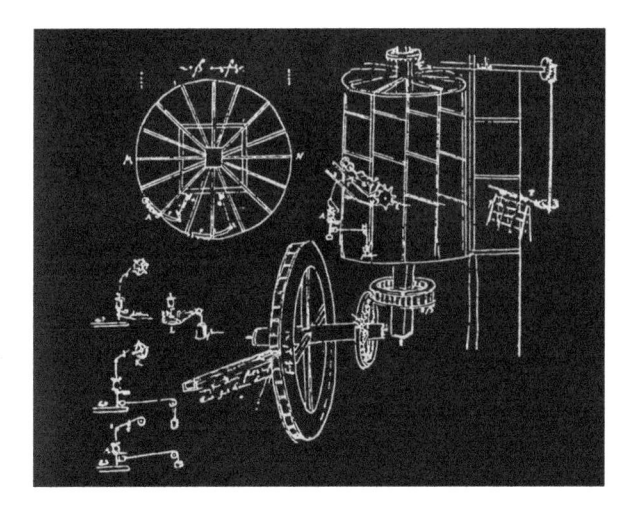

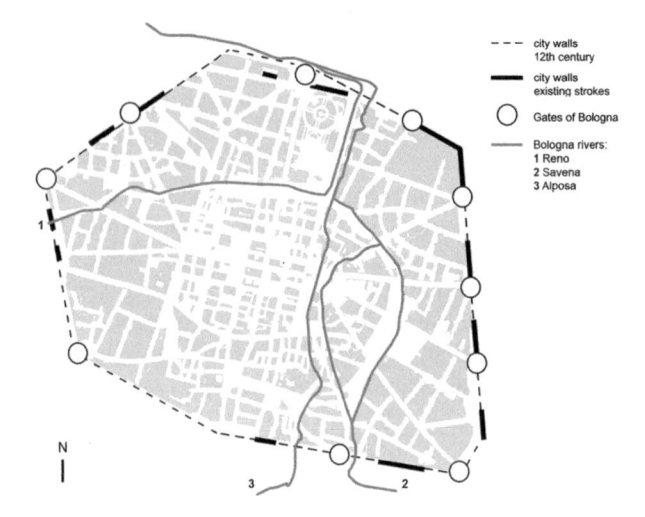

Fig. 2 Rivers of Bologna

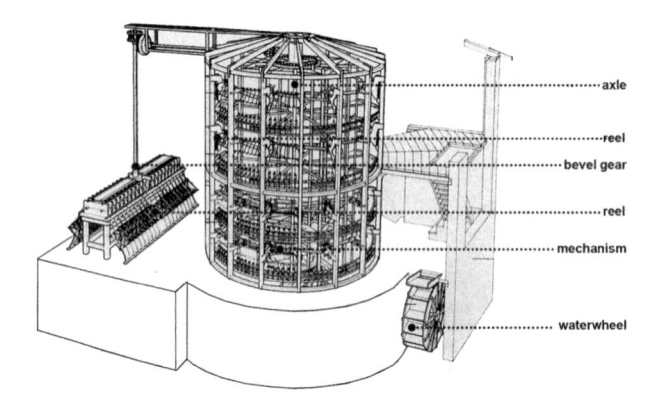

Fig. 3 Axonometric silk mill view

Interestingly, it is precisely here and not at the "historical" seat of spinning that in 1341 a spinner from Lucca introduced here an innovative method of production, which would constitute a breaking point in silk production in Bologna, namely the first mechanical spinning wheel powered by an overshot water wheel. Its structure was quite well adapted to the form and energy demand offered by the hydraulic and architectonic systems of Bologna. The dimensions of these machines, much smaller than that of blade water wheels, made it possible to install them in cellars of buildings thus guaranteeing continuity and greater power of water flow.

Fig. 4 Silk mill 3D mesh model

Fig. 5 Silk mill details

3 Silk Production in Bologna: Buildings and Factories

During the period of greatest expansion (XVIII century) there were 119 silk mills within city walls [6], all of them powered by 353 water wheels of 1 or 2 HP. They were all powered by water flowing to cellars in the whole block [5]. The mills employed an overall number of 600–700 adult men at spinning wheels and about 1500 children at winders, producing 300000 silk thread annually. The process of production took place in circumstances which could easily be transformed from habitable spaces into factories: sluices and canals ran under street level, the wheels were installed in cellars, the machines (spinning wheels and winders) were distributed along the vertical axis on various floors of the buildings.

The mills were not isolated cells of the city texture. On the contrary, they constituted an essential element of production network, which involved energy sources, machinery, engines, men, streets and products. The mill brought together functions of living quarters with those of a factory. Working space was organized vertically, in accordance with the structure of the spinning wheels and throwing machines. Water powered engines installed in cellars propelled machines for silk throwing located in the central part of the building as well as mechanical winders placed in attics. The habitable space and that of the factory were clearly divided: on the former space—light, fire of the kitchen, clean and clear well water; in the latter —darkness, cold, noise and black water for industrial use. The entrance hall in the centre both united and separated the two realities: family life and work organized on the principles of the factory system of labour. Historians of industrial revolution confirm that the Bologna silk mills (spinning wheels—throwing machines provided with winders propelled by water wheels) meet exactly the requirements that transform a manufacture into a factory.

4 Silk in Bologna: Water and Technology

La Silk is a natural fiber produced by silkworms in the form of cocoons. Since antiquity silk thread several hundred of meters long has been unwound from the cocoon in a process called "reeling". At this stage the necessary tools were just a basin and a reel. A worker forms a single thread from several cocoons bathed in hot water which then another woman winds on a reel, forming a skein. Gradually the process of reeling concentrated on great spinning wheels where tens of women worked machines that were improved and perfected all the time. But before being woven the threads are worked again: one or more of them are strengthened and shrunk by repeated throwings. This operation bears precisely that name, silk throwing, traditionally done by hand with small domestic tools like the spindle.

The great heyday of silk industry in Bologna began with the changing of silk throwing methods. The innovative mill was a hand operated machine which made it possible to draw and throw threads from hundreds of reels. In 1340 Bolognini di Borghesano di Bonaventura from Lucca was granted permission to construct a spinning wheel attaching the spinning element of a water wheel. Thus a complex machine came into existence, one that performed various stages of silk working at the same time. This machine brought prosperity to the economy of Bologna between 1200 and 1700, particularly because it introduced diverse innovative technologies. By applying the overshot water wheel to the Lucca prototype, the small spinning wheels installed in rooms were transformed into machinery that took up 3–4 floors of a building. The mechanism of an overshot wheel functioned largely in the following way: water flowed into the compartments which emptied when the wheel made half a circle downwards; the rotating wheel transmitted circular movement onto a horizontal axis to which one or two vertical wheels made of wood or iron were attached.

The technological innovations can be expressed in a few fundamental points.

First of all, the introduction of sluices [4], small tubes that directed water from canals in order to bring it exclusively to silk mils. They made it possible to draw water from the main duct to a smaller one important mainly near the cellars of buildings where compartment water wheels rotated. As opposed to grain mills, where water flowed from below, in the silk mills it arrived from above. The mechanism that propelled mills initiated with the wheel, which rotated anticlockwise and moved the central tree which rotated a great number of threads onto reels and spools. Silk mills did not need water in great quantities to rotate their wheels. That is why it was forced to flow from canals to the mills by sluices. Moreover, the introduction of sewer tunnels made it possible to reuse waste water. The movement of spinning wheels was obtained by vertical compartment water wheels of low potency installed in the cellar of a building. To rotate them it was enough to provide a relatively small quantity of water by means of sluices. This particular solution ensures a more even, thinner and resistant yarn in comparison to that obtained with a hand operated or animal operated spinning wheel. It moreover gave the owner the possibility to distribute the spinning on various levels of the building where tens of workers could work at the same time, unlike the four of five with the two of them propelling the machine.

The functioning model of a silk mill housed at the museum of industrial heritage in Bologna is 34 m high and 23 m of diameter and documents its technological aspect. It does not reflect a philological documentation, on the contrary, it is a mixed type model with two supports. The upper part contains a throwing machine "*a bacchette*", while the lower part "*a guindani*" is complete with the water wheel and hollow tiles (tavellas) for the winder.

5 Silk in Bologna: Graphic Documentation

The existent graphic documentation is rather scarce, containing moreover, few detailed descriptions or the architectonic structure or the mechanisms of silk mills. In Bologna the technology underlying them was kept in strictest secrecy. It was feared the diffusion of the secret to other cities would create dangerous and unwanted competition. However, despite grave punishments provided for violators of this law, the water spinning wheel system of Bologna reached the north of Italy at the beginning of the XVI century and then was successively exported abroad. For reasons of secrecy mentioned above images of the silk mill "*alla Bolognese*" that survived to our day are but few, patchy and do not tell us much. The oldest representation can be found in the manuscript of "Trattato dell'arte della seta" (Treatise on the Art of Silk Making) from Florence of 1487—a copy of an earlier version from the end of the XIV century. Extant are also some designs of Leonardo and Heinrick Schickhardt (1599). However, the best description of the functioning of the precious machinery was provided by Vittorio Zonca. In his treatise "Novo teatro di macchine et edifici", published in Padua in 1607, he depicts the machinery in its whole complexity, wary all the time not to provide elements which would reveal exactly its construction. In the treatise there are two tables of drawings and four with text. From the pages of the treatise we learn that *"the factory with the water powered spinning wheel is beautiful, even marvelous, because one can see wheels, spindles, little cogwheels and other wooden objects in transverse position, their length and diagonal placement so clearly that the eye of the beholder loses itself inside it and makes one ponder how human genius could have comprehended such a variety of things, of movement in one and in opposite direction, propelled by one wheel that is inanimate..."*. The famous Encyclopédie of Diderot and D'Alambert published in the XVIII century also provided a detailed description of the silk mill with drawings of the whole and its details.

The construction of the model exhibited in the museum of industrial heritage is based on scarce graphic documentation which define the essential parts of the main carrying structure (supports, *poste*, stanchions), of the movement transmission systems (big cogwheels linked to the water wheel, mobile internal part complete with its *serpi*, star shaped wheels, belts operating each spindle by friction) as well as those of movement transforming (crankshaft system) and the operating parts (spools, thread guides, *guindani*, rods) [2].

Acknowledgments We hereby thank the Museo del Patrimonio Industriale - Istituzione Bologna Musei - Comune di Bologna (http://www.museibologna.it/patrimonioindustriale) for allowing us to take photographs of the functioning mechanical model of waterwheel and the mechanical spinning wheel.

References

1. Bloch, M.: Avvento e conquiste del mulino ad acqua. ID, Lavoro e tecnica nel Medioevo **73–110**, 201–219 (1970)
2. Ceccarelli, M., Cigola, M.: On the evolution of graphical representation of gears in mechanisms, transmissions and applications. In: Lovasz, E. and Corves, B. (eds.) History of Mechanisms and Machine Science, vol. 3. Springer (2011)
3. Guenzi, A., Poni, C.: Sinergia di due innovazioni. Chiaviche e mulini da seta a Bologna. In: Quaderni storici **XXII**, 64 (1987)
4. Guenzi, A.: Acqua e industria a Bologna in antico regime. Giappichelli (1993)
5. Pini, A.I.: Energia e industria tra Savena e Reno: i mulini idraulici tra XI e XV secolo. In: Tecnica e società nell'Italia dei secoli XII-XVI. Centro italiano di studi di storia e d'arte (1987)
6. Pini, A.I.: L'acqua nella città medievale. Canali e mulini a Bologna tra XI e XV secolo. Scuolaofficina **IX**(1) (1990)
7. Zanotti, A.: Il sistema delle acque a Bologna dal XIII al XIX secolo. Compositori (2000)

Robots in History: Legends and Prototypes from Ancient Times to the Industrial Revolution

A. Gasparetto

Abstract Even in ancient times the idea of "robots", intended as artificial beings that could substitute real individuals to carry out heavy and repetitive tasks, flourished and led to the birth of many legends. In addition, several ingenious inventors, belonging to different epochs and civilizations, designed and built prototypes of what we can define "robots". In this paper, we sketch a brief history of Robotics throughout the centuries, from ancient times to the Industrial Revolution (18th century), describing the most interesting legends and the most relevant examples of robot prototypes that were designed and/or built.

Keywords Robotics · Automata · History · Prototypes · Legends

1 Introduction

The idea of artificial beings that could substitute real individuals, especially to carry out heavy tasks, dates back to ancient times, with the birth of many legends, among different cultures and civilizations. For many centuries, however, artificial devices could not be built because the technology was not developed enough. Only after the Renaissance, a significant progress in technology, especially in the methodologies for metal processing, enabled the most ingenious inventors to build autonomous mechanical devices.

There are not so many books or journal papers devoted to the history of Robotics. A historical overview may be found in the work by Ceccarelli [6], or in the book by Rosheim [20], or within the book by Rossi et al. [22]. Some information on history of Robotics may be found in Wikipedia [12].

This paper presents a synthetic history of Robotics from ancient times to the 18th century, i.e. to the time when the progress of the Industrial Revolution set the basis

---

A. Gasparetto (✉)
Polytechnical Department of Engineering and Architecture,
University of Udine, Udine, Italy
e-mail: alessandro.gasparetto@uniud.it

for the widespread diffusion of automatic devices in the industrial environment, thus ending the "amateur" era of Robotics.

The paper is organized as follows. In Sect. 2, legends created by ancient cultures, concerning artificial individuals or beings, are described. In Sect. 3, the most relevant designs and prototypes of artificial beings conceived up to the 18th century are described and discussed. In Sect. 4 conclusions are drawn.

2 Legends

Many legends about "robots" flourished in different epochs and civilizations. In ancient Egypt, the priests used to make animated statues to communicate the divine will to the people. In the Canadian far north and in the Western Greenland the Inuit legends tell us of *Tupilaq*: anthropomorphic beings that could be created by a sorcerer to hunt and kill enemies.

In the Greek mythology there are several tales concerning "robots". The most famous one is the myth of Pygmalion, a sculptor from Cyprus who made an ivory statue of a girl and called it Galatea (Fig. 1). The statue was so realistic that he fell in love with it and wished it could become a real woman. Eventually, Aphrodite, the goddess of love, fulfilled his wish and Galatea came to life. The most beautiful

Fig. 1 "Pygmalion priant Vénus d'animer sa statue", by Jean-Baptiste Regnault (1786). Salon des Nobles du Château de Versailles (*source* Wikimedia Commons)

version of this myth can be found in the *Metamorphoses* by Ovid [17]. In another myth, Cadmus killed a dragon and buried its teeth, which suddenly turned into armed soldiers. With these men, Cadmus founded the city of Thebes and became its first king [13]. Moreover, according to classical mythology, Hephaestus, the crooked God of metals, created mechanical servants of various types and complexity, varying from intelligent golden young ladies to more utilitarian three-legged tables that could move autonomously [14]. Another example of a robot that can be found in Greek legends is Talos, a gigantic bronze automaton built by Daedalus to defend the island of Crete by throwing huge boulders against an attacking fleet [14]. Some other legends tell us that in Thebes were built statues able to speak and move their arms, while in Heliopolis it was possible to admire statues that could come down from their pedestals.

The idea of anthropomorphic beings was not considered in the first centuries of the Middle Ages, mainly due to religious taboos. Only in the 13th century the idea of some sort of "robot" came back to life. According to a legend, the philosopher Roger Bacon (1214–1294), a Franciscan monk, built a brass head that could talk and answer to any possible question [5]. A similar head was owned by St. Albertus Magnus (1200–1280), philosopher and Doctor of the Catholic Church.

In the 16th century, it was the turn of the alchemists: they gave "recipes" to build not only the head, but even the entire body of an artificial anthropomorphic being, which they named *Homunculus* ("small man") [11]. The recipe provided by Paracelsus (Philippus Aureolus Theophrastus Bombastus von Hohenheim, 1493–1541), the famous Swiss physician and alchemist, contributed to the birth of the legend of Faust.

The idea of artificial anthropomorphic beings is present also in the Jewish culture: such beings are called *Golem*. The most famous Golem legend dates back to the 16th century: the rabbi of Prague, Judah Loew ben Bezalel, created the Golem to save Prague Jews from antisemitic attacks. He went at night to the Vltava River with two of his assistants, and with the clay of the river banks they shaped a human figure. At the end of the rite, the rabbi imprinted on the front of the creature a holy word, which gave life to the inanimate matter. The Golem was able to read people's minds, thus identifying those who wanted to harm the Jews [15].

3 Designs and Prototypes

In addition to legends, the ancient Greek world left us some examples of technical designs of automata. According to Diogenes [8], the first design of a "robot" can be considered the one proposed by the Greek mathematician Archytas of Tarentum (428–347 BC), who conceived the first ever known artificial machine: a bird-shaped device named "*the pigeon*" that could fly under the propulsion of steam.

Diogenes [8] also tells us of Ctesibius (285–222 BC), a Greek mathematician living in Alexandria, who made several inventions, among which some automata intended to "please the eye and the ear": a blackbird that sang by means of a water flow, as well as human figures which could drink and move. Ctesibius also invented a water organ that is considered the precursor of the modern pipe organ, and a water clock that for many centuries (before Huygens' pendulum) was the most accurate clock in the world (Fig. 2). Some time later, Hero of Alexandria, a mathematician and engineer who lived in the first century AD, in his work *Automata* [4] described some devices that, using the steam produced by a rudimentary boiler or the force generated by a water jet, made statues move, as in Fig. 3.

In ancient China, some inventors designed ingenious automata. An interesting example is the *Cattle Machine* (described in [24]), which was used for transportation of heavy loads. In 1088 Su Song (1020–1101 AD) designed the *Cosmic Engine*, a clock tower featuring mechanical dummies that could sound the hours, ring the gongs and the bells of the device [16].

In Arabic civilization, the inventor Al Jazaari (1136–1206 AD) designed several automatic devices, including musical automata powered by water and humanoid robots that could play music and entertain during the royal parties (Fig. 4). The interesting feature of these robots lies in the fact they could be "programmed" by adjusting the cams that implemented the percussion on the drums, thus allowing to play different melodies [2]. In Japan, until the end of 18th century, some automata were built for the amusement of rich people. The most popular version was the "tea maidservant", a humanoid device which could move and offer a cup of tea (Fig. 5).

In order to see the first projects of automata in the Western world, one must wait until the latest centuries of the Middle Ages. Villard de Honnecourt, a French artist and engineer of the early 13th century, drafted several automata: however, they were more sketches than designs in a modern sense.

Fig. 2 The water clock by Ctesibius [18]

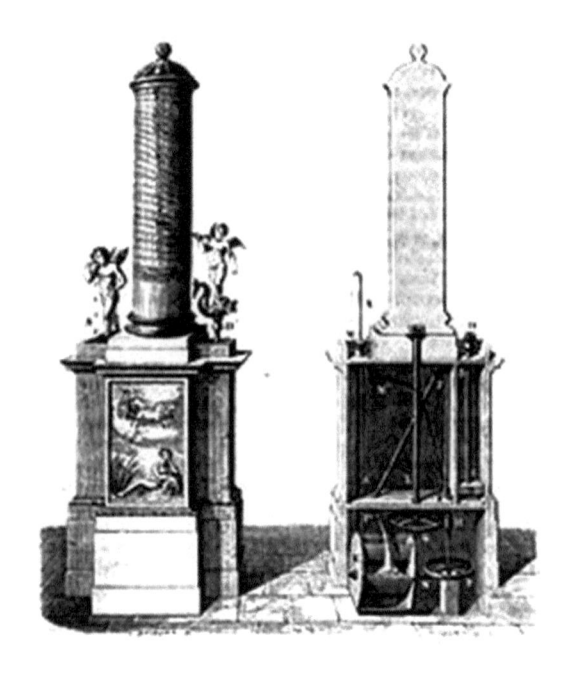

Fig. 3 An automaton by Hero of Alexandria (Reconstruction made by Giovanni Battista Aleotti 1589) [1]

Fig. 4 An automaton by Al Jazaari [2]

Examples of "robots" in the late Middle Ages were the moving figures built to enrich the towers and the clocks of the churches. An extraordinary example is the mechanical cock found in the clock of the Strasbourg Cathedral, built in the 14th

Fig. 5 Modern reconstruction of the Japanese tea maidservant automaton [20]

century (Fig. 6): at the stroke of noon, the cock came out of a door, opened his beak, showed his tongue, beat its wings and uttered three times a loud crowing.

In the Renaissance, the Italian engineer Giovanni da Fontana (1395?–1454?) in his *Bellicorum Instrumentorum Liber* ("Book on war devices") described and drew several automata to be used to throw arrows or bombs against enemies [7].

The first documented project of a programmable humanoid robot was made by Leonardo da Vinci (1452–1519) around 1495. Some notes by Leonardo, rediscovered in the 1950s, contain detailed drawings for the construction of a mechanic knight, who was apparently able to stand up, wave his arms and move his head and jaw (Fig. 7). It is not known whether the project, which was probably based on anatomical research culminating in the famous Vitruvian Man, was ever implemented by Leonardo. More details may be found in [21].

Later in the 16th century, the German mathematician and astronomer Regiomontanus (Johannes Mueller von Koenigsberg, 1436–1476) designed and built an artificial iron fly and an iron eagle that could fly [23]. In England, a wooden beetle capable of flying was designed by the astronomer John Dee (1527–1608) [10].

Before the 18th century, the limited development of mechanical technology did not allow to build real robotic devices, so most of the "robots" found their place only on paper, not in the real world. However, over the centuries the techniques for metal processing greatly improved, thus increasing the accuracy of precision mechanisms. Such progress in the mechanical technology enabled inventors to build real anthropomorphic automata.

Fig. 6 The mechanical cock that used to be in the clock of Strasbourg's cathedral, now at the Musée des arts décoratifs de la ville de Strasbourg (*source* Wikimedia Commons)

In 1738 Jacques de Vaucanson (1709–1782) built a mechanical duck that could eat, drink, squawk and eject the waste products (Fig. 8). The duck wings, alone, contained a large number of moving parts: about 400. De Vaucanson also built two musicians robots: a drum player and a flute player. The latter could play the flute through a system of pipes that carried to the instrument an air flow coming from a bellows. More details about de Vaucanson's life and work may be found in [9].

Between 1770 and 1773 two Swiss brothers, Pierre (1721–1790) and Henri-Louis (1752–1791) Jaquet-Droz, built three impressive humanoid automata: a scribe, a draftsman and a musician (Fig. 9). The scribe was able to write letters with up to 40 characters and, by replacing a disk, could write different types of text. The draftsman could perform various types of drawings: from the portrait of Louis XV to a ship with its sails. The musician android was a 16 years old girl, who

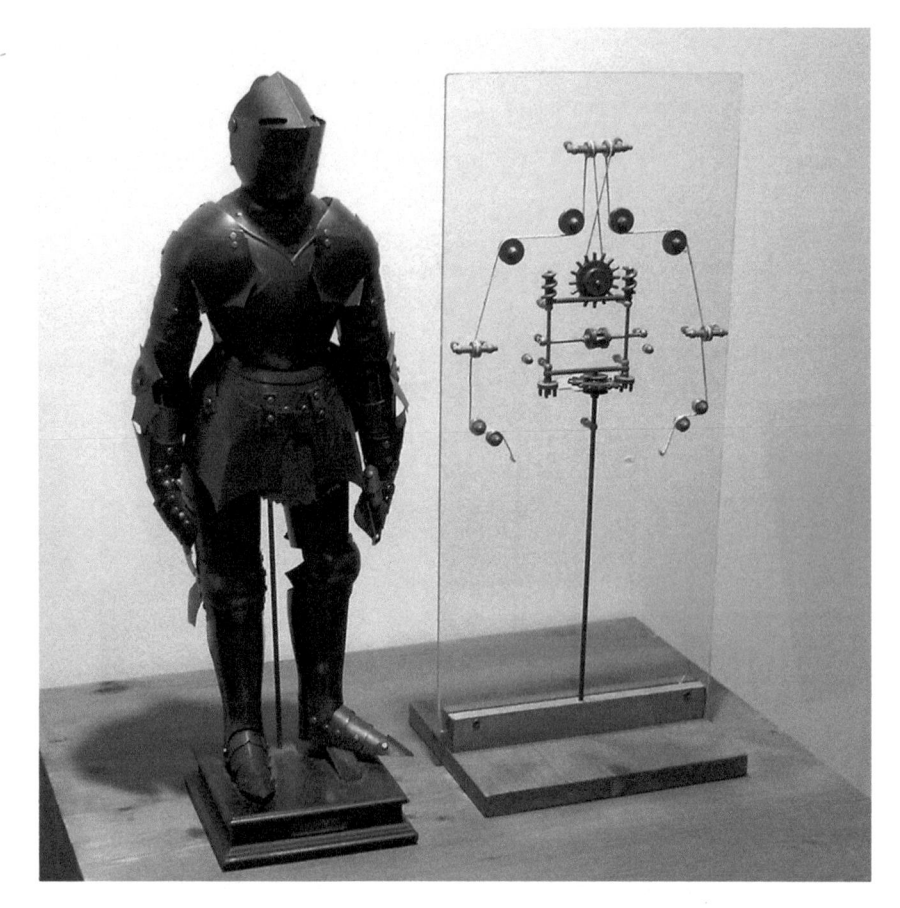

Fig. 7 Reproduction of Leonardo Da Vinci's knight and his internal mechanisms at the "Leonardo da Vinci. Mensch—Erfinder—Genie" exhibit, Berlin 2005 (*source* Wikimedia Commons)

was able to play an organ mimicking even the pauses and bowing at the end of the exhibition. These robots are still working and are located in the Musée d'Art and d'Histoire in Neuchâtel (Switzerland). More details about the Jaquet-Droz brothers and their automata may be found in [19].

The 18th century therefore appears as a time of extraordinary results in the field of Robotics. Until that time, all the automata were made by ingenious inventors mainly for pleasure. The following century saw the consolidation of the Industrial Revolution, and all the energies of inventors were devoted to the design and the development of devices that could increase the industrial production (for instance, the programmable loom invented by Joseph-Marie Jacquard in 1801), or to the mass production of automata, such as the thousands of clockwork singing birds that were built and exported all around the world by many small family-based automata makers living in Paris in the years 1848–1914 (this period was therefore named by

Fig. 8 The original De Vaucanson's duck at Musée des Automates de Grenoble, destroyed in a fire in 1879 (*source* Wikimedia Commons)

Bailly "the golden age of automata") [3]. In this sense, the 18th century marks the end of the amateur times of Robotics: from then on, the industrial and commercial needs became the main driving forces for the development in that field.

Later on, the 20th century can be considered the time when electronics starts to be widely employed in Robotics. Besides that, in that century scientists begin to think of building machines that could not only perform a mechanical repetition of the movements of a human being, but could also make decisions, being provided with memory and criteria for choosing. In other words, the history of humanoid Robotics in the 20th century is connected with the history of Electronics, Computer Science and Artificial Intelligence.

Fig. 9 Jaquet-Droz's automata at the Musée d'Art et d'Histoire de Neuchâtel (*source* Wikimedia Commons)

4 Conclusions

This paper dealt with a brief history of Robotics, from ancient times to the Industrial Revolution (18th century). The idea of artificial beings that could substitute real individuals to carry out heavy or repetitive tasks dates back to ancient times. Many legends among different civilizations, from ancient Egypt to the Renaissance, have been described. Thanks to the development of mechanical technologies, inventors could build devices that one could previously just imagine: several of such devices have been described in this paper. The historical outlook extends up to the 18th century, i.e. to the end of what we may consider the amateur era of Robotics.

References

1. Aleotti, G.B.: Gli artifitiosi et curiosi moti spirituali di Herrone tradotti da M. Gio. Battista Aleotti d'Argenta. Aggiontovi dal medesimo quattro theoremi non men belli e curiosi de gli altri. Et il modo con che si fa artificiosamente salir un canale d'acqua viva o morta in cima d'ogni alta torre. Ferrara, Baldini Editore (1589)

2. Al-Jazaarí: The Book of Knowledge of Ingenious Mechanical Devices: Kitáb fí ma'rifat al-hiyal al-handasiyya. Translated by Donald Routledge Hill. Dordrecht, Reidel Publishing Company (1974)
3. Bailly, C.: Automata: The Golden Age: 1848–1914. Robert Hale Editor, London (2003)
4. Baldi, B.: Degli automati, overo màcchine se moventi. Translation of Automata by Hero made by Bernardino Baldi. Florence, Girolamo Porro Editore (1589)
5. Butler, E.M.: The Myth of the Magus. Cambridge University Press (1948)
6. Ceccarelli, M.: A Historical Perspective of Robotics Toward the Future. Fuji International Journal of Robotics and Mechatronics **13**(3), 299–313 (2001)
7. Clagett, M.: The Life and Works of Giovanni Fontana. Annali dell'Istituto e museo di storia della scienza di Firenze **1**(1), 5–28 (1976)
8. Diogenes L.: Lives of Eminent Philosophers. Translated by R.D. Hicks, Cambridge. Harvard University Press (1972—First published 1925)
9. Doyon, A., Liaigre, L.: Jacques Vaucanson, mécanicien de genie. PUF, Paris (1966)
10. Fell-Smith, C.: John Dee (1527–1608). Constable & Co., Publishers, London (1909)
11. Grafton, A.: Natural Particulars: Nature and the Disciplines in Renaissance Europe. MIT Press, Cambridge (1999)
12. http://en.wikipedia.org/wiki/History_of_robots
13. Kerényi, C.: The Heroes of the Greeks. Thames and Hudson, London (1959)
14. Kerényi, C.: The Gods of the Greeks. Thames and Hudson, London (1979)
15. Moshe I.: Golem: Jewish Magical and Mystical Traditions on the Artificial Anthropoid. State University of New York Press, Albany (1990)
16. Needham, J.: Science and Civilization in China. Caves Books, Taipei (1986)
17. Ovid: The Metamorphoses. Translated by A. Mandelbaum. Harcourt Brace, London (1993)
18. Perrault, C.: Les Dix Livres d'Architecture de Vitruve. Jean Baptiste Coignard, Paris (1673)
19. Perregaux, C.: Les Jaquet-Droz et leurs automates. Wolfrath & Sperlé Neuchâtel (1906)
20. Rosheim, M.E.: Robot Evolution. Wiley (1994)
21. Rosheim, M.E.: Leonardo's Lost Robots. Springer (2006)
22. Rossi, C., Russo, F., Russo, F.: Ancient Engineers' Inventions: Precursors of the Present. Springer (2009)
23. Wilkins, J.: The mathematical and philosophical works of the Right Rev. John Wilkins, late lord bishop of Chester. Vernor and Hood, London (1802)
24. Yan, H-S.: A Design of Ancient China's Cattle Machine. In: Proceedings of the 10th World Congress on Theory of Machines and Mechanisms, pp. 57–62 (1999)

The First Hundred Years of Mechanism Science at RWTH Aachen University

B. Corves

Abstract The dedicated field of mechanism theory at RWTH Aachen University has been established institutionally in 1959 under Walther Meyer zur Capellen. However, this does not mark the real beginning of mechanism theory at RWTH Aachen University. In fact mechanism theory was a teaching and research topic already from the very beginning of RWTH Aachen University in 1870 influenced by Franz Reuleaux and later in the late Twenties of the last century especially through the activities of Kurt Rauh. In the following it will be shown how mechanism science in Aachen has always played a role during the first hundred years of RWTH Aachen University.

Keywords Mechanism theory · RWTH Aachen university · Franz Reuleaux · Kurt Rauh · Walther Meyer zur Capellen · Mechanism models

1 Introduction

On October 1870 the "Königliche Rheinisch-Westphälische Polytechnische Schule" (Royal Rhenish-Westphalian Polytechnical School) which was later to become RWTH Aachen University opened its doors for 32 teachers and 223 students. In fact as mentioned in [1] the Prussian state government promoted the regional development towards "*a gateway to progressive technologies*" in the Aachen area in many ways, e.g. by giving up-and-coming companies modern machines modelled on the English example, however, with the provision that they make them accessible to other companies for examination and assessment. The establishment of the subsequent RWTH Aachen University in 1870 in Aachen was also one of these regional "structural development" acts.

B. Corves (✉)
Department of Mechanism Theory and Machine Dynamics, RWTH Aachen University, Aachen, Germany
e-mail: corves@igm.rwth-aachen.de

It was Franz Reuleaux (Fig. 1) as a renowned academician, kinematician and engineer [2, 3] who was involved in the planning and development of the new engineering institution. As a member of a commission advising on the structure and curriculum of the new institution he drew up a draft budget and as argued by [3] he obviously used his influence in making Adolf von Gizycki (*1834, +1891, Fig. 2), who had previously been an assistant and lecturer at the "Königliche Gewerbeinstitut" (Royal Industrial Institute) in Berlin under Franz Reuleaux's auspices, one of just three other professors of still young mechanical engineering at RWTH Aachen University.

Fig. 1 Franz Reulaux (1829–1905) Newspaper print on his 70th birthday

Fig. 2 Adolf von Gizycki, Professor of "Kinematics Theory, (Reuleaux System)" at Aachen Technical University from 1870 to 1891

2 Early Mechanism Theory in Aachen

From the start of mechanical engineering teaching in Aachen it was von Gizycki who represented the fields of *"Descriptive Machine Theory"*, *"Theoretical Machine Theory"* and *"Theory of Kinematics (Reuleaux System)"*. Thus from the very beginning of RWTH Aachen University in 1870 "modern" Reuleaux type mechanism theory was therefore at home in Aachen [1]. Another indication to the fact that this *"Theoretical Kinematics (Reuleaux System)"* course of von Gyzycki played an important role in the mechanical engineering education of RWTH Aachen University can be seen from the fact that there exists an inventory notice of the "Kinematic Collection" of RWTH Aachen as displayed in an exhibition catalogue of an 1880 "Industry and Art Exhibition" which was held in Düsseldorf in 1880 [4]. The participation of RWTH Aachen University was organized by a group of professors with von Gizycki as part of the organizing committee representing mechanical engineering. In section "III Maschinenkunde" (Machine Engineering) of the exhibition catalogue eighty kinematic models which were on display for the "Industry and Art Exhibition" are listed under the heading "Kinematik". According to this list almost all of these kinematic models can be traced back to Reuleaux models as described in the famous Voigt catalogue [5] or in the "Skizzenbuch der angewandten Kinematik" (Sketchbook of Applied Kinematics) [6]. Only the group of "Steuerrudergetriebe" (helm or rudder mechanisms) with models ascribed to Reed, Scott and Sinclair, Rogers, William, Reuleaux and Steel the author could not trace back to one of the above mentioned Reuleaux models. According to [4] the complete Kinematic Collection contained 195 demonstration models and according to the above mentioned catalogue the models have been produced in the workshops of the Royal Industrial Academy of Berlin, costing 15200 Mark in total, see also Fig. 6 in [3].

Comparing the complete list of Aachen Reuleaux Models with the list of models from the famous Voigt catalogue [5] reveals that the Aachen collection represents the majority of Reuleaux models. Figure 3 shows an excerpt of one page of this

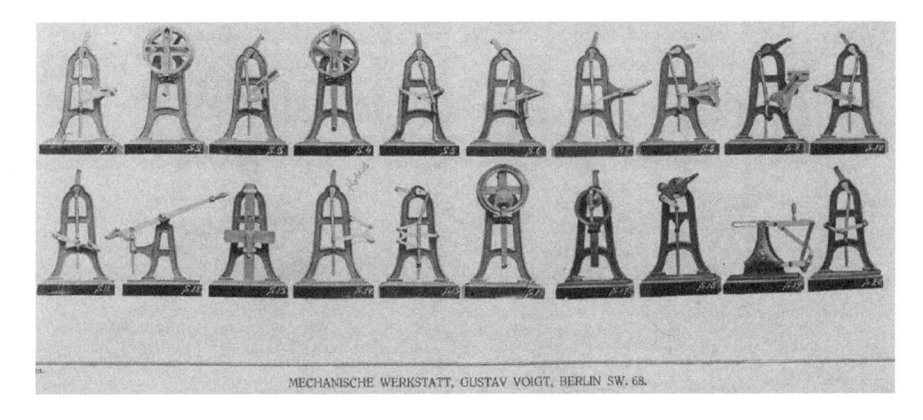

Fig. 3 Excerpt of the "Gustav Voigt Catalogue" from 1907 with models S1 to S20 [5]

catalogue with the first twenty models of totally 39 models just related to straight line motion. All of these twenty models shown in Fig. 3 can be traced to have been part of the Aachen collection. A detailed description and drawings of the Reuleaux models including calculation formulas can be found in [6].

Unfortunately as already stated in [3] these in engineering terms very valuable Reuleaux models, discussed and described in detail by Moon [2] and Corves and Braune [7] apparently disappeared from RWTH Aachen. In [3] the authors also reported that in 1994 in the University of Hannover almost twenty Reuleaux models were found in an unused storage basement awaiting a clear-out, and were rescued at the last hour by one of the authors from being finally scrapped. A complete overview with descriptions, pictures, videos and interactive animations can be found through the following web link: http://www.dmg-lib.org/dmglib/main/portal.jsp?mainNaviState=browsen.set.viewer&id=73046.

Six of these Hannover models shown in Fig. 4 are related to straight line guidance models and a comparison of the above mentioned exhibition catalogue reveals that they can be assumed to be nearly identical to models that have been part of the large Reuleaux model collection at RWTH Aachen University.

As also discussed in [3] there is no direct evidence of the "Theoretical Kinematics (Reuleaux System)" teaching of Prof. Gizycki in Aachen but there exists a very good handwritten school exercise book of a student named Edw.

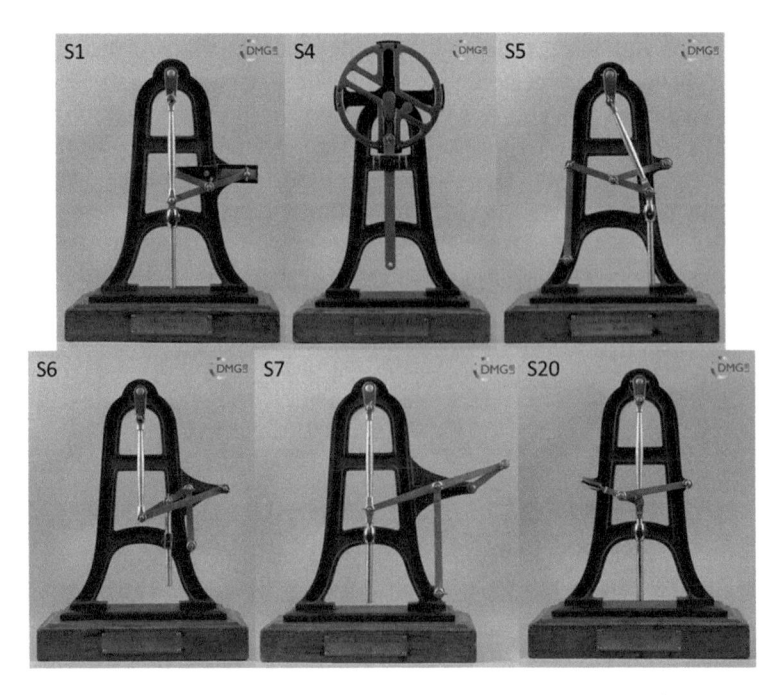

Fig. 4 Six of the Hannover models which can be assumed to be nearly identical to models that have been part of the large Reuleaux model collection at RWTH Aachen University

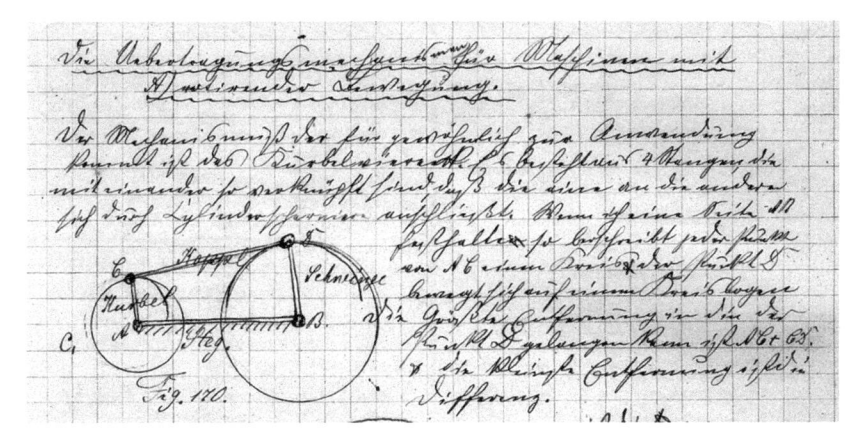

Fig. 5 Lecture notes of the student Edw. Neufang

Neufang from the winter term 1876/77 of the lecture on "Descriptive Machine Theory" by Prof. Gizycki. Figure 5 shows an excerpt of this document which shows one of various mechanisms sections that clearly convey its Reuleaux basis [8].

As already mentioned in [3] the section titled *"Transmission mechanisms with rotating motion"* begins in the same way as a present day beginner's course in mechanism theory could begin: *"The mechanism usually applied is the four-bar linkage."* In fact the sketch of a four bar mechanism shown in Fig. 5 looks very much like *the* demonstration model of Reuleaux for a core theses of his "constraints theory", the generation of different mechanisms from a single chain by selecting different links of the chain as a fixed frame while retaining all relative movements. In fact in his major work on "The Kinematics of Machinery" [9] the four bar kinematic chain and the derived crank and rocker mechanism play an important role (Fig. 6). Whereas the latter is depicted three times in Reauleaux's figures 11, 21 and 180 the four bar chain is shown even five time in figures 10, 20, 153, 179 and 190.

As already described in [3] towards the beginning of the twentieth century the subject of "Kinematics" appears to have gradually lost significance as an independent field in Aachen. Nevertheless in 1909 Eugen Essich finished a dissertation with the title *"About control mechanisms with rolling cams"* [10] which appears to be the first "mechanisms" dissertation after 1899, when German Institutions of Higher Technical Education received the right to confer doctorates.

Fig. 6 Figures 179 and 180 from Reuleaux's "The Kinematics of Machinery" [9]

3 Renewal of Mechanism Theory in Aachen by Kurt Rauh

During the time of World War I and shortly afterwards the author was not able to trace special activity in the field of mechanism theory. Next evidence of renewed activity is a letter of the Institute Director of the Agricultural College in Bonn, Prof. Dr.-Ing. Karl Vormfelde, dated April 17th, 1928 to the Prussian Minister for Agriculture, Domains and Forestry in Berlin, which marks the re-beginning of mechanism theory by a dedicated lecturer in persona Kurt Rauh at RWTH Aachen University with the following words (see also [11]):

> In order that agricultural engineering gets a firm footing also at RWTH Aachen University which is in the interest of German agriculture, I have caused my assistant, Dr.-Ing. Rauh to habilitate in Aachen with the topic "Kinematics and Agricultural Engineering". RWTH Aachen University will agree and I would therefore respectfully want to ask to approve to the habilitation of Dr.-Ing. Rauh in such a way that he will hold for 1 day each week during the winter term a 2 or 3 h lecture in Aachen. The work of the Agricultural Machinery Institute in Bonn-Poppelsdorf will not be negatively influenced by this habilitation. The scientific work and progress of Dr. Rauh will ensure an honourable success.

Kurt Rauh had received his engineering education at the then "Technische Hochschule Dresden" (TH Dresden) now Technical University of Dresden in

November 1921. Immediately afterwards he started working as a design engineer for mechanical handling devices for a mill manufacturer and machine company in Dresden but already in spring 1922 he started a new job as scientific assistant at the "Institut für Maschinentechnologie" (Institute for Machine Technology) of Prof. Rudolf Hundhausen again at TH Dresden. But also this engagement did not last long possibly due to economic circumstances and after a short engagement in Leipzig he moved to Bonn in 1923, where he started as the head of innovation and design offices and experimental department of the company F. Soennecken. In 1924 he also became director of the company's patent office. In 1926 Kurt Rauh quit his job in order to pursue a job as scientific assistant and later as scientific chief assistant at the "Landwirtschaftliche Hochschule Bonn-Poppelsdorf" (Higher Agricultural College in Bonn-Poppelsdorf) [11].

At that time the Agricultural College in Bonn did not have the permission to promote candidates with the "Dr.-Ing"-title which is why Kurt Rauh submitted his dissertation [12] to the former Technische Hochschule Hannover (TH Hannover), today Leibniz University Hannover, where he received his Dr.-Ing.-title on July 12th, 1927. In his dissertation entitled "Investigation and development of a periodic mechanism with forward and reverse and acceleration-free operation," he describes a task still valid today in many applications even if increasingly servo-controlled drive systems are used for this purpose today [13]. Interestingly, throughout the history of mechanism theory in Aachen in the second half of the last century, mechanisms with temporarily synchronous velocity have been a special topic. At least five research reports [14–18] which originated under the auspices of Walter Meyer zur Capellen, the successor of Kurt Rauh as head of mechanism theory in Aachen, can be mentioned here among other publications. Still the successor of Walter Meyer zur Capellen, Günter Dittrich kept the topic alive as can be seen from the dissertations of Shi and Leusch [19, 20].

So only shortly after finishing his dissertation and receiving the doctoral degree as Dr.-Ing., Kurt Rauh had also received the habilitation at RWTH Aachen University and the position of a private lecturer while keeping his position as Chief-Assistant at the Agricultural Machinery Institute in Bonn-Poppelsdorf until 1934. More details of Kurt Rauh's life and contribution to mechanism theory can be found in [11, 13]. His influence and success in his teaching and research activities is not only related to just mechanism theory [21, 22] but also to agricultural engineering [23, 24] and patent issues [25]. A list of his main works can be found in [11]. Figure 7 shows an excerpt of [21] which clearly shows the Reuleaux influence when compared to Figs. 5 and 7.

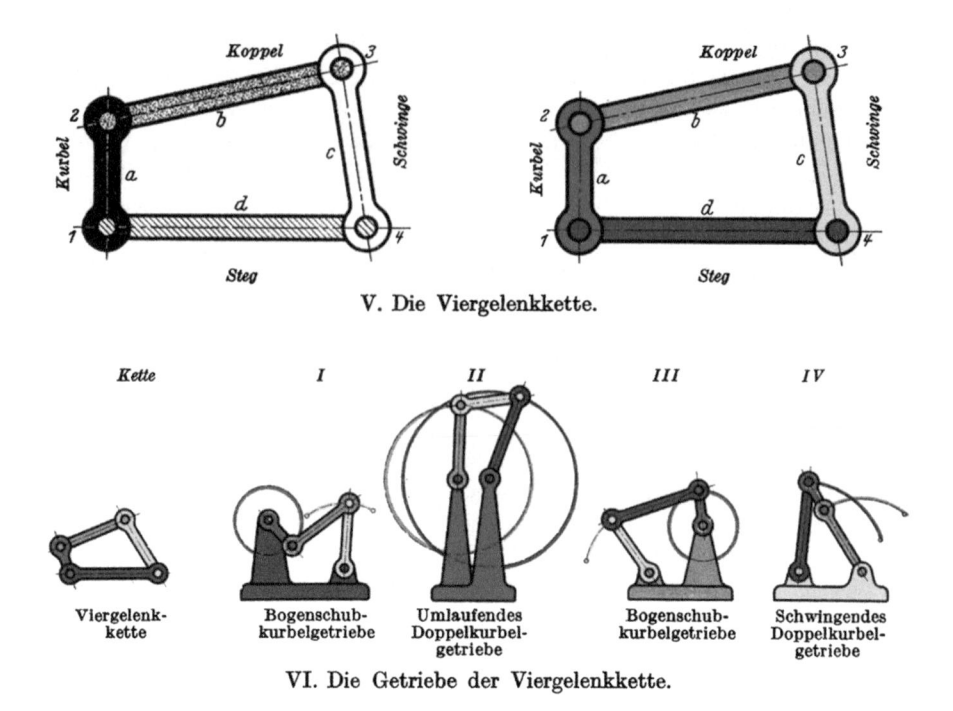

Fig. 7 Figures from [21] showing the influence of the Reuleaux concept of kinematic chains

4 Mechanism Theory and Dynamics of Machines Under Walther Meyer Zur Capellen

After completing his engineering studies at TH Darmstadt in 1926, Walther Meyer zur Capellen (Fig. 8) decided to continue his academic education by taking up an assistant post also in Darmstadt, as well as a second degree in mathematics, physics, philosophy, psychology and education, leading to a graduate teaching qualification in 1928 [26]. On 17th February 1932, his dissertation topic "Method on approximated solutions of Eigenwert problems focusing on vibration problems" [27] earned him his doctorate and the title Dr.-Ing. under Prof. Dr. A. Walther at the Institute for Practical Mathematics at the Technical University Darmstadt.

Already in the 1930s Walther Meyer zur Capellen had made himself a name by publishing several mechanism theory papers of mathematical character. Upon attaining his doctorate he started on October 1st, 1932 as a lecturer in the position of "Baurat im technischen Schuldienst" (government building officer in technical school service) at the "Technische Staatslehranstalt für Maschinenwesen" (Technical State Teaching Institution for Mechanical Engineering), which is today known as the "Fachhochschule Aachen" (University of Applied Science Aachen).

Following the sudden death of Kurt Rauh in 1952, Walther Meyer zur Capellen was appointed as a lecturer for mechanism theory at RWTH Aachen University and

Fig. 8 Walther Meyer zur
Capellen (1902–1985)

in 1956 he achieved his habilitation for the field of mechanism theory. During this time he did not only publish works on mechanism theory and machine dynamics, but also mathematical contributions such as the 180 page work "Leitfaden der Nomographie" (Guide to Nomography), [28]. Shortly after the beginning of his teaching activities at RWTH Aachen University he began to build up a collection of mechanism models. His first paper where he initially refers to mechanism models from his collection [29] was published in an uncommon publication called "Fette, Seifen, Anstrichmittel, Die Ernährungsindustrie" (Grease, Soaps, Coating Material, The Nutrition Industry) in 1957. This shows that Walther Meyer zur Capellen always had the practical application of his research results in mind, which cannot be taken for granted looking at his very mathematical career. Most of the models described in this paper still exist in the mechanism collection at IGM. The first models consisted of ASTRALON©, a polymethylmethacrylate (PMMA). Later models were produced from acrylic glass so they could be displayed during lectures and presentations using overhead-projectors. Except for the picture on the bottom left, which shows a current CAD-model, Fig. 9 shows models of mechanisms described in [29]. The mechanism shown here and titled "Zykloidenlenker" (cycloid linkage) resembles very much the well known Tchebychev linkage but the dimensions are different, see also [30].

At the beginning of 1959 the "Chair for Mechanism Theory" was established and on January 13, 1959 Walther Meyer zur Capellen was named Associate Professor. The preceding calling negotiations mainly dealt with the provision of jobs for research assistants as well as "*a commitment to provide appropriate resources for the establishment of a workshop and a laboratory as well as to equip the other rooms of the building*". In 1964 the name of the Chair was changed to "Department of Mechanism Theory". Since then Walter Meyer zur Capellen was the head of the department as full professor, whose teaching and research area was enlarged by the field of "Dynamics of Machines" in 1966. He remained professor and director of the department until his retirement in 1970.

Abb. 9a. Modell des Zykloidenlenkers

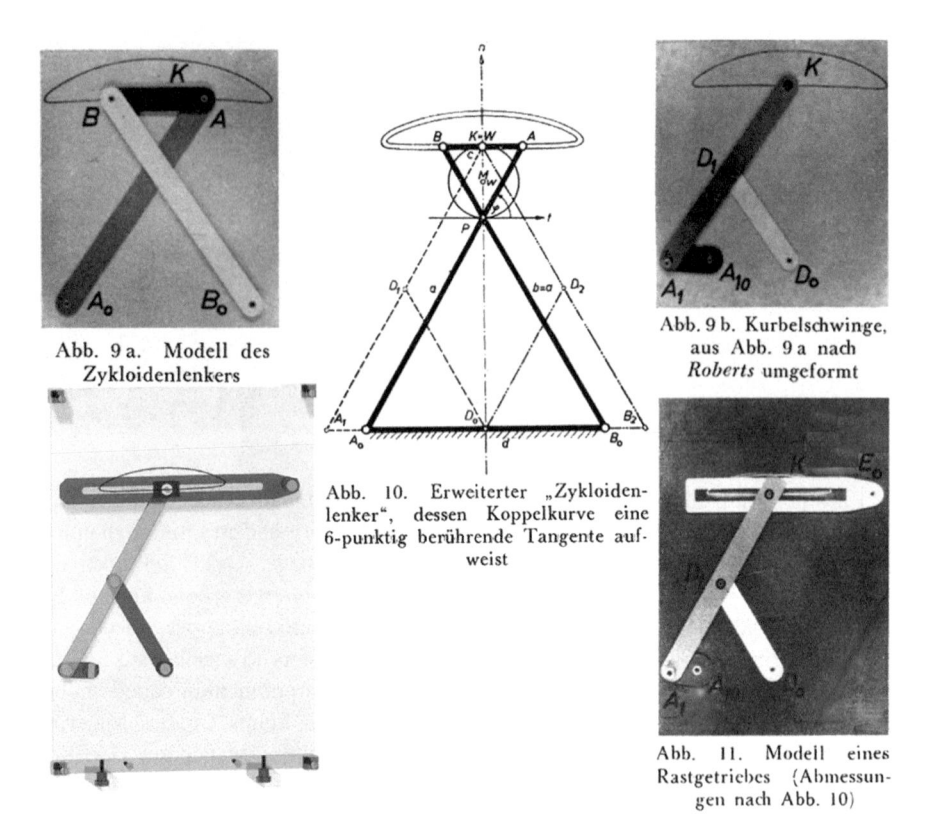

Abb. 9b. Kurbelschwinge, aus Abb. 9a nach *Roberts* umgeformt

Abb. 10. Erweiterter „Zykloidenlenker", dessen Koppelkurve eine 6-punktig berührende Tangente aufweist

Abb. 11. Modell eines Rastgetriebes (Abmessungen nach Abb. 10)

Fig. 9 Mechanism models from [29]

Besides the aforementioned ASTRALON©-models, metal models were created during Walter Meyer zur Capellen's time as head of the department. These included planar and spherical mechanisms. On the upper left side of Fig. 10 a planar dwell mechanism from [31] is shown together with the according CAD-model on the lower left. On the upper right side a photo of a spherical Geneva Wheel mechanism from [32] and its CAD-model on the lower right is shown.

A good overview of Meyer zur Capellen's research work is provided by an article in the 1965 Yearbook on Research issued by the State of Northrhine-Westphalia [33] and an issue dedicated to the Department of Mechanism Theory and Dynamics of Machines of the Industrieanzeiger (Industrial Journal) [34], where Walter Meyer zur Capellen also acted as editor.

The latter publication especially refers to the new research and teaching facilities after moving the department into the new faculty building in 1962 with a large workshop on the fifth and sixth floor and ample space to display the growing mechanism model collection (Fig. 11).

Until 2012, when the department moved out due to refurbishment of the building the showcases with the different mechanism models on display still served this

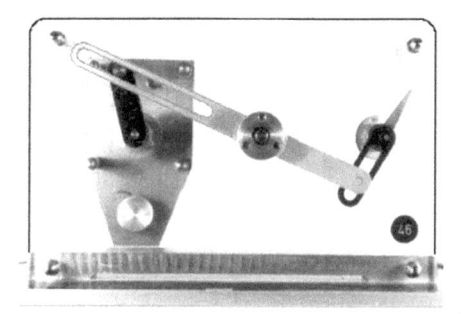

Fig. 10 Metal Mechanism models from [31, 32] and their CAD-Models

Fig. 11 IGM mechanism model collection

purpose and have been the basis of several activities, such as Mechanism descriptions [30], Mechanism encyclopedia [35] and the project DMG-Lib [36].

As stated in [34] Walther Meyer zur Capellen had published about 110 books and journal papers with additional 80 publications (including dissertations and research reports) of his employees. At the same time he gave 45 invited lectures, 16 of those on a four week trip to the United States in October 1966. In fact, about 40 years ago Frank Erskine Crossley mentioned three "German Missionaries" [37] that travelled to the United States in the 1950s and 1960s. Their names were Kurt Hain, Rudolf Beyer and Walther Meyer zur Capellen. In order to describe the influence those three men had on the research and teaching of Mechanism Theory in the United States, Crossley wrote the following: *"Just as in the sixth century Irish saints brought Christianity to the Germans, so now the Germans brought their kinematic science to the U.S."*

In 1961 Walther Meyer zur Capellen took over from Kurt Hain the tradition to systematically register and classify relevant literature in the field of Mechanism Theory by using corresponding overviews [38], which he continued until 1969. After that this task has been carried on by Günter Dittrich, who was already engaged with this duty as a co-author since 1963 [39], and who also authored the last overview in 1997 [40].

The best way to appreciate the work and services of Walther Meyer zur Capellen is through the words of Herwart Opitz [34]. According to him Walther Meyer zur Capellen was *"significantly involved in combining theoretical knowledge with experiences from the field to issue guidelines for engineers working in the field. (…) Mechanism Theory as an independent field of teaching and research has its roots in Germany and is connected to names such as Reuleaux, Burmester, Alt and R. Müller. His reputation as the leading German scientist in the field of Mechanism Theory goes far beyond the borders of our country."*

5   Conclusions

This paper indicates the deep roots of Mechanism Theory at RWTH Aachen University and especially at the Department of Mechanism Theory and Dynamics of Machines. From the beginning of the "Königliche Rheinisch-Westphälische Polytechnische Schule" (Royal Rhenish-Westphalian Polytechnical School) in 1870 mechanism theory has played an important role. Started under the influence of Franz Reuleaux especially the activities of Kurt Rauh and Walther Meyer zur Capellen, were aimed at a systematic approach towards the synthesis and selection of mechanisms of different kind.

References

1. Gast, P. (ed.): Die Technische Hochschule zu Aachen 1870–1920, RWTH Aachen, 1920—Library, Signature Ka21 = 3
2. Moon, F.C.: The Machines of Leonardo Da Vinci and Franz Reuleaux. Springer, Dordrecht (2007)
3. Braune, R., Corves, B.: Franz Reuleaux and Rhenish Roots of Mechanism Theory. In: Explorations in the History of Machines and Mechanisms, Proceedings of HMM 2012, May

7–May 11, VU University, Amsterdam (Series: History of Machines and Machine Science), pp. 21–38. Springer, Dordrecht, Heidelberg, New York, London. ISBN: 978-94-007-4131-7 (2012)
4. RWTH Aachen: Verzeichnis der Gegenstände, welche ausgestellt sind von der Königlichen-Rheinisch-Westfälischen Technischen Hochschule zu AACHEN nebst Mitteilungen über die Anstalt (Gewerbe- und Kunstausstellung in Düsseldorf 1880), Ausstellungskatalog, F. N. Palm in Aachen (1880). http://darwin.bth.rwth-aachen.de/opus/volltexte/2009/2792/
5. Voigt, G.: Kinematische Modelle nach Professor Reuleaux. Berlin (1907) http://ebooks.library.cornell.edu/k/kmoddl/toc_voigt1.html and http://ebooks.library.cornell.edu/k/kmoddl/toc_voigt2.html
6. Reuleaux, F. Althof, J., La Baume, Boy, H., Lührs, Schittke, M.: Skizzenbuch der angewandten Kinematik, Eine Sammlung von Mechanismen nach den Vorträgen des geheimen Regierungsrath Professor Reuleaux. Verein „Hütte", Berlin (1880)
7. Corves, B., Braune, R.: International geschätzt, verschollen und wiedergefunden. Getriebemodelle von Franz Reuleaux In: Das materielle Model. Wilhelm Fink, Paderborn (2014)
8. Neufang, E.: Skript zur Vorlesung "Beschreibende Maschinenlehre" von Prof. v. Gizycki im WS 1876/77, RWTH Aachen – Signatur Ff 2117/HBRA
9. Reuleaux, F.: The Kinematics of Machinery, Outlines of a Theory of Machines. Macmillan and Co, London (1876)
10. Essich, E.: Über Steuerungsgetriebe mit Wälzhebeln, Dissertation bei Prof. Langer, RWTH Aachen, 1909
11. Corves, B.: Kurt Rauh (1897–1952) In: M. Ceccarelli (ed.), Distinguished Figures in Mechanism and Machine Science, History of Mechanism and Machine Science, vol. 26. Springer Science + Business Media, Dordrecht (2014). ISBN: 978-94-017-8946-2, S./Art.: 231–262
12. Rauh, K.: Untersuchung und Weiterentwicklung der Getriebe mit periodischem Hin- und Rücklauf und beschleunigungsfreiem Arbeitsgang. Diss. TH Hannover. Rhenania-Verlag G. m.b.H., Bonn (1927)
13. Corves, B.: Kurt Rauh: Saxon Mechanism Theory Brought to Aachen. In: Explorations in the History of Machines and Mechanisms, Proceedings of HMM 2012, May 7–May 11, VU University, Amsterdam. ISBN: 978-94-007-4131-7, S.: 107–122
14. Meyer zur Capellen, W.: Eine Getriebegruppe mit stationärem Geschwindigkeitsverlauf. Forschungsbericht Nr. 606 des Landes Nordrhein-Westfalen, Köln, Opladen: Westdeutscher Verlag 1958
15. Meyer zur Capellen, W., Lehn, F: Kinematische Kenngrößen der ebenen elliptischen und der räumlichen Schleifen. Forschungsbericht Nr. 1593 des Landes Nordrhein-Westfalen, Köln, Opladen: Westdeutscher Verlag 1966
16. Meyer zur Capellen, W., Schreiber, E.: Getriebe mit stationärem Geschwindigkeitsverlauf. Forschungsbericht Nr. 1851 des Landes Nordrhein-Westfalen, Köln, Opladen: Westdeutscher Verlag 1967
17. Meyer zur Capellen, W., Schreiber, E.: Raumgetriebe mit stationärem Geschwindigkeitsverlauf. Forschungsbericht Nr. 2096 des Landes Nordrhein-Westfalen, Köln, Opladen: Westdeutscher Verlag 1970
18. Meyer zur Capellen, W., Willkommen, W.W.: Sphärische Viergelenkgetriebe als Proportionalgetriebe. Forschungsbericht Nr. 2386 des Landes Nordrhein-Westfalen, Köln, Opladen: Westdeutscher Verlag 1970
19. Shi, Z.: Maßsynthese von Getriebekombinationen aus ebenen oder sphärischen Doppelkurbeln und anderen Getrieben zur Verwirklichung bereichsweise vorgegebener Übersetzungsverhältnisse. Dissertation RWTH Aachen 1982

20. Leusch, G.: Ein Beitrag zur Systematisierung und Maßsynthese ebener un-gleichmäßig übersetzender Getriebe mit bereichsweise annähernd konstanter Übersetzung. Dissertation RWTH Aachen, VDI Fortschrittbericht Reihe 1, Nr. 248, Düsseldorf: VDI-Verlag 1995
21. Rauh, K.: Praktische Getriebelehre. Verlag von Julius Springer, Erster Band. Berlin (1931)
22. Rauh, K.: Praktische Getriebelehre. Zweiter Band. Berlin: Verlag von Julius Springer (1939)
23. Rauh, K.: Entwicklungslinien im Landmaschinenbau. Girardet, Essen (1949)
24. Rauh, K.: Bearbeitungsmaschinen – angewandte Getriebelehre. VDI-Z **92**(33), 917–922 (1950)
25. Rauh, K.: Ich melde meine Erfindung selbst an: Vorlesungen über Patentlehre an der Technischen Hochschule Aachen zusammengefasst zu fünf Rundfunkvorträgen (1932 übertragen vom Westdeutschen Rundfunk). Erweiterte Ausgabe als Handschrift gedr., Köln: Borowsky, 1935
26. Corves, B.: Kurt Hain und Walter Meyer zur Capellen: A View from Aachen at Two Shapers of German Mechanism Theory. In: MeTrApp (Workshop on Mechanisms, Transmissions and Applications) in Timisoara, 06.10. 10.10.2011. Springer, 2012. ISBN: 978-94-007-2726-7. ISSN: 2211-0984, S. 15-35
27. Meyer zur Capellen, W.: Methode zur angenäherten Lösung von Eigenwertproblemen mit Anwendungen auf Schwingungsprobleme. (Dissertation), Ann. d. Phys. 5. Folge (1931) Nr. 6, S. 299–352
28. Meyer zur Capellen, W.: Leitfaden der Nomographie. Berlin, Göttingen, Heidelberg: Springer (1953)
29. Meyer zur Capellen, W.: Ueber gleichwertige periodische Getriebe, Z.: Fette, Seifen, Anstrichmittel, Die Ernährungsindustrie 59 (1957) Nr. 4, S. 257–266
30. Dittrich, G.; et al.: IGM-Getriebesammlung. Aachen 1995
31. Meyer zur Capellen, W.: Die Kreuzschleife als Rastgetriebe. Techn. Mitteilungen HdT 56 (1963) Nr. 7, S. 1/8
32. Meyer zur Capellen, W.: Sphärische Maltesergetriebe. T. M. 54 (1961) Nr. 7, S. 239–244
33. Meyer zur Capellen, W.: Das Institut für Getriebelehre an der Rheinisch-Westfälischen Technischen Hochschule Aachen. In: Der Ministerpräsident d. Landes NRW. Landesamt f. Forschung, Jahrb. 1965
34. Meyer zur Capellen, W. (Hrsg.): Industrie-Anzeiger, Antrieb, Getriebe, Steuerung, 89. Jg. (1967) Nr.34
35. Corves, B., Niemeyer, J.: Das IGM-Getriebelexikon als Instrument der Wissensverarbeitung in der Getriebetechnik. Proceedings of the IX. International Conference on the Theory of Machines and Mechanism, Aug 31–Sept. 2, 2004, Liberec, Czech Republic
36. Brix, T., Döring, U., Corves, B., Modler, K.H.: DMG-Lib: the "Digital Mechanism and Gear Library". Project. In: Proceedings of the 12th World Congress in Mechanism and Machine Science, June 18–21, 2007, Besancon, France
37. Crossley, F.R.E.: Recollections From Forty Years of Teaching Mechanisms. Trans. ASME: Jl. of Mechanism, Transmissions, and Automation in Design, vol. 110 (1988), 3, pp. 232–242
38. Meyer zur Capellen, W.: Ungleichförmig übersetzende Getriebe (Schrifttumsübersicht 1961). VDI-Zeitschrift. Düsseldorf: VDI-Verl., Bd. 104 (1962) Nr. 6, S. 301–03
39. Meyer zur Capellen, W., Dittrich, G.: Ungleichförmig übersetzende Getriebe (Jahresübersicht 1962). VDI-Zeitschrift 105 (1963) Nr. 6, S. 273–77
40. Dittrich, G., Stolle, G.: Ungleichmäßig übersetzende Getriebe (Jahresübersicht 1997). VDI-Zeitschrift 140 (1998), Nr. II (Special Antriebstechnik), S. 70–77

Mock-Up of an Eighteenth-Century Oil Mill via Rapid-Prototyping

J.C. Montes, R. López-García, R. Dorado-Vicente and F.J. Trujillo

Abstract Information technologies and Computer Aided Engineering (CAE) programs open new ways for conservation, documentation and study of out of date mechanisms. Resulting virtual animations and numerical simulations help to understand and recover our industrial heritage. Currently, get a mock-up from a virtual model is easier and cheaper thanks to low cost rapid prototyping techniques. These scaled models give a more vivid experience of a mechanism and its performance. This paper describes a methodology to obtain a mock-up of an XVIII century oil mill, part of the historical heritage of Fondón (Almeria, Spain). We digitalized the mill components in CAD software using field measurements, got a CNC code to print them on a Fused Deposition Modelling (FDM) machine, and printed and assembled the mill.

Keywords History of machines and mechanisms · 3D printing · Geometric modelling

1 Introduction

The rapid industrial development (mainly in the last centuries) accelerates the obsolescence of industrial equipment and machines. Their suppression leads to loss of technical knowledge [1].

J.C. Montes · R. López-García · R. Dorado-Vicente (✉) · F.J. Trujillo
Department of Mechanical and Mining Engineering, Universidad de Jaén,
Jaén, Spain
e-mail: rdorado@ujaen.es

J.C. Montes
e-mail: jcmg0002@red.ujaen.es

R. López-García
e-mail: rlgarcia@ujaen.es

F.J. Trujillo
e-mail: fvilches@ujaen.es

Since its birth around 1950, industrial archaeology aims to protect and understand the technical and industrial knowledge developed in the past. This relatively new area of research highlights the cultural and scientific interest in disused industrial equipment and systems. The reader interested in the evolution of this research area may find interesting the description of Sanchiz [2].

The restoration of old industrial artefacts contributes to understand the evolution of our current society, and can attract new economic activities based on the industrial heritage.

Moreover, technical knowledge conservation is key to reduce the effort of new engineering developments. Laroche and Bernard [3] point out that "...conserving those objects ... can become a source of knowledge for anticipating our "future" and for creating objects of tomorrow...".

Nevertheless, as Hudson [4] indicates, the cost of keeping a single machine makes unviable to save all of them. On the other hand, document old industrial equipment is an affordable task, which can be performed via detailed reports and photographs.

The development of computers and the information technologies contributes to the aforementioned documentation task because of computer capabilities to store and manage large amounts of data in digital format. Digital photos, 3D scans, animations and videos of the industrial remains are possible today [5].

Several researchers claim that Computer-Aided Design (CAD) programs can benefit industrial archaeology [6]. These programs lead to virtual models (digital mock-ups [7]) that can simulate the behaviour of ancient industrial systems as different works portray [3, 8–10]. Moreover, CAD software helps to generate and store technical drawings providing a more relevant documentation for engineers.

Nowadays, rapid prototyping techniques based on additive-manufacturing techniques simplifies the process to obtain a physical model from a CAD file. Physical models offer a more vivid experience of an object or system [1], and enable us to test it in real conditions.

The Fused Deposition Modelling (FDM [11]) is a low-cost additive technique that produces parts from a plastic filament (mainly ABS and PLA).

This paper describes how to obtain a scaled physical model of an ancient industrial mechanism using an FDM 3D printer. We reproduce the grinding system of an eighteenth century oil-mill at the town of Fondón (Almeria, Spain). The oil-mill is described in Sect. 2. Section 3 explains the digitalization process using field measurements and CAD software (Solidworks). In Sect. 4, we describe the FDM printer, and the printing process: save parts in STL format, slice each part and get CNC code, and finally print all parts. The main conclusions are drawn in Sect. 5.

2 Olive Oil and Oil Mills

2.1 Olive Oil

Olive oil is a key product in Mediterranean countries and because of that its production is an interesting topic. Spain is the largest producer of olive oil in the world (60 % of world production), and Andalusia (southern Spain) is the largest producer in Spain (80 % of Spanish production) [12].

The production of olive oil has evolved throughout history, so that different disused devices can be found in Andalusia. Virtual models and mock-ups help to preserve the technical and cultural knowledge related to those devices.

2.2 Mill and Press

The mechanical operations to produce olive oil are: grind the olives to break the cells containing the olive oil, and press or centrifuge the resulting olive paste to separate the olive oil from the solid raw material.

Different systems for grinding and pressing olives have been developed throughout history. Before the Romans, mortars were used for olive crushing, while pressing was done with the help of woven mats. Romans developed the previous processes. They gave us the *trapetum*, two hemispheric millstones driven manually, and different presses based on screws.

The aforementioned processes did not change too much till the XVIII century, when new rotating stone wheels driven by animals and different type of presses (such as the beam press and quintal) were developed.

Since XVIII century to XX century, traditional techniques powered by animals were used to grind olives and press the olive paste (Fig. 1). These techniques are currently obsolete because of their low profitability.

Nowadays, continuous production techniques use metal drums instead of cylindrical or conical millstones (Fig. 2), and a centrifuge system for separating the olive paste in 2 or 3 phases. The interested reader can find a more detailed description of the evolution of olive oil production processes in the work of Kapellakis et al. 13].

We focus on the digitalization and physical reproduction of an eighteenth-century grinding system with conical rollers. The following Sect. 2.3 shows the main characteristics of the oil mill studied.

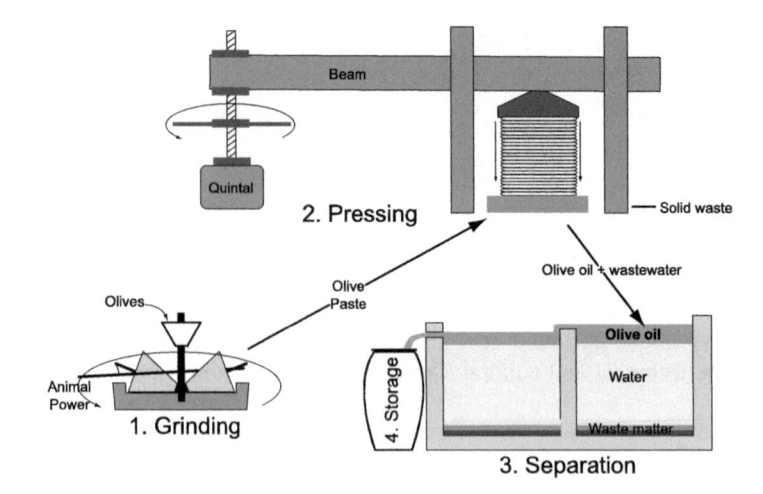

Fig. 1 Olive oil extraction in the XVIII century. In this example the grinding system is a set of conic roller stones and the olive paste is pressed by a quintal and beam mechanism

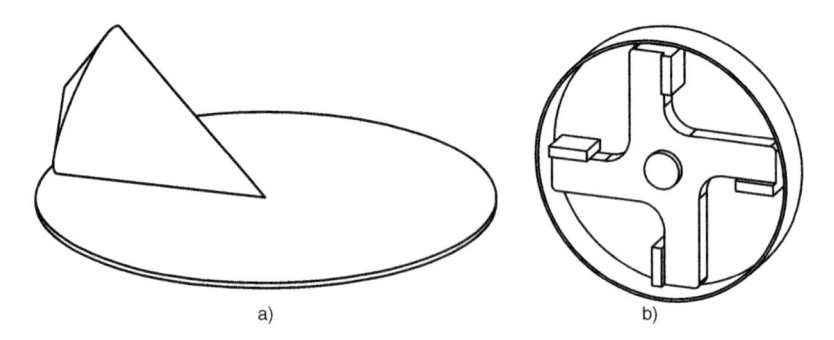

Fig. 2 Mill scheme: **a** traditional conical millstone, **b** current metal roller

2.3 Oil Mill in Fondón

The oil mill studied is located in the municipality of Fondón (South of Sierra Nevada, Spain), about 850 m above sea level on the Andarax river.

Fondón town's population is estimated at 1,000 inhabitants, and the economy is based on intensive farming of vegetables and olives. The area has a Mediterranean climate with high temperatures and sporadic rainfalls.

The oil mill was built in the late eighteenth century and has been owned by the Aguilera family for three generations. The original mill consisted of two conical granite rollers driven by animals (the development of the conical surface matches the mill floor or base what improves the crushing process and decreases the passive resistances of the machine), a system of beam press and quintal (dimension of the

Fig. 3 Current status of the oil mill studied

Fig. 4 Illustration of the studied grinder: basic components and operation

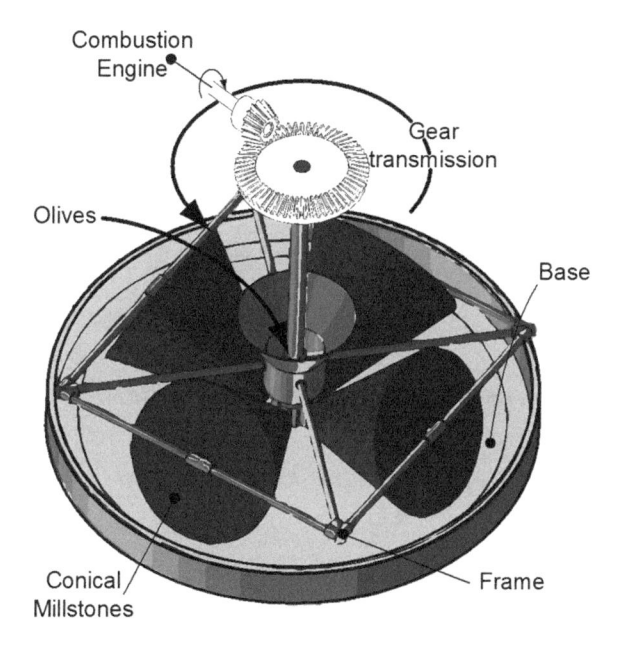

beam 15 m), and a system of tailings ponds where olive oil was separated from the pressed olive paste (Fig. 1).

Later, the system was modified: a feed system consisted of an Archimedean screw driven by an electric motor was added, the original grinding system was substituted by four granite mill stones driven by a combustion engine, and the beam and quintal by a hydraulic press.

This work is focussed on the four-roller grinder. Figure 3 shows details of the current status of that mechanism. On the other hand, Fig. 4 gives a more clear description of the grinder's components and its operation.

3 CAD Modelling

This section is devoted to describe the main steps to model the previous oil mill (Sect. 2, Fig. 4) using Solidworks, a well-known Computer-Aided Design (CAD) software. Solidworks allows parametric modelling of three-dimensional objects, assembling and animating mechanisms, generating exploded views and so on. Resulting models can be exported in different formats.

Modelling steps in Solidworks:

- Measure the main elements of the mill using a measuring tape. Draw each part using the Solidworks tools "sketch" and "operations" (Fig. 5).
- Open "assembly" module and place each part using mating relationships (Fig. 6).

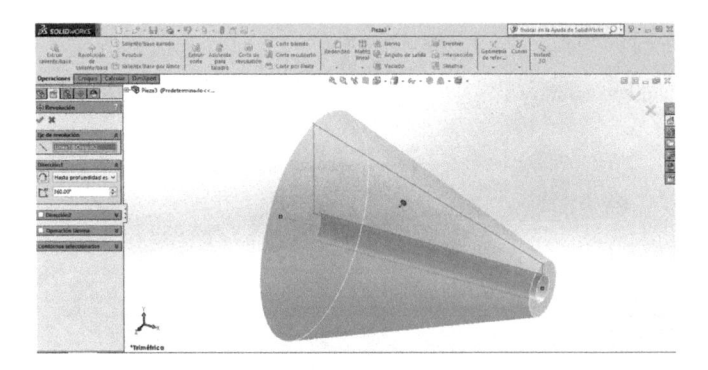

Fig. 5 Solidworks tools: "scheme" and "operation"

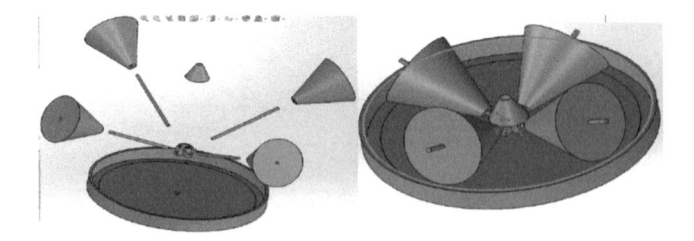

Fig. 6 Virtual oil mill assembly via solidworks

Fig. 7 "Export STL" options in solidworks

- Generate drawings and/or export parts in other formats. Parts should be discretized and exported in STL (Stereo Lithography) format; this step helps to obtain the CNC code. Figure 7 shows the STL exportation dialog box in Solidworks.

4 Printing Process

The main goal of this work is to obtain a scaled model of an oil mill consisting of 4 conical rollers on a circular base (3.5 m diameter). We scale rollers and base models (scale 3/50), and print each element using a FDM printer. A frame of steel wire joins rollers and chute.

4.1 FDM Printing

Fused deposition modelling (FDM) is a rapid prototyping technique based on additive manufacturing where a plastic filament (usually PLA or ABS) is melted and extruded through a small orifice. The extruder, controlled by a CNC program, fills layer by layer the volume of a part. FDM technology was developed in the early nineties and nowadays it has become the most commonly used Additive Manufacturing (AM) technique.

According to Gibson et al. [14], FDM technology has the following advantages:

- Wide range of materials.
- FDM provides stronger polymer parts than other AM processes.
- Cost of FDM machines and materials are lower than other AM processes.

On the other hand, inertia limits the extruder acceleration, and therefore the operation speed is low compared to other manufacturing processes.

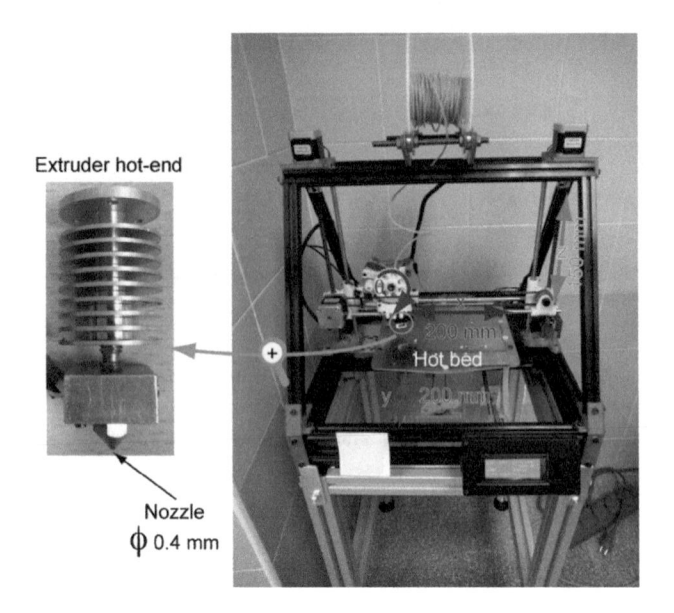

Fig. 8 BCN3D main characteristics

4.2 FDM Printer Characteristics

We use a BCN3D machine (Fig. 8). This printer is based on RepRap project, which provides hardware designs and software to control different printer variants [15]. These machines can be assembled with minimum cost and tools. An assembled machine, similar to the BCN3D model, costs around 900 €.

4.3 CNC Code Generation

The mechanical and geometrical characteristics of each printed component depend on the printing parameters and the extruder path defined in a CNC code.

There are different programs that provide automatically the previous G-code from an STL version of a CAD file. The idea is to slice the piece in different flat layers, and then define the extruder path to fill each layer. There are different fill patterns: direction parallel, spiral, honeycomb etc. Feed speed and layer height are defined in the G-code and influence the processing time and the printing quality.

Because of its simple interface and its accessibility (it is an open-source application); we use Repetier-Host to define the main printing parameters and to get the G-code program. Figure 9 shows the software interface and its main tools: visualization, object placement, slice and G-code generation, and manual control of the printer.

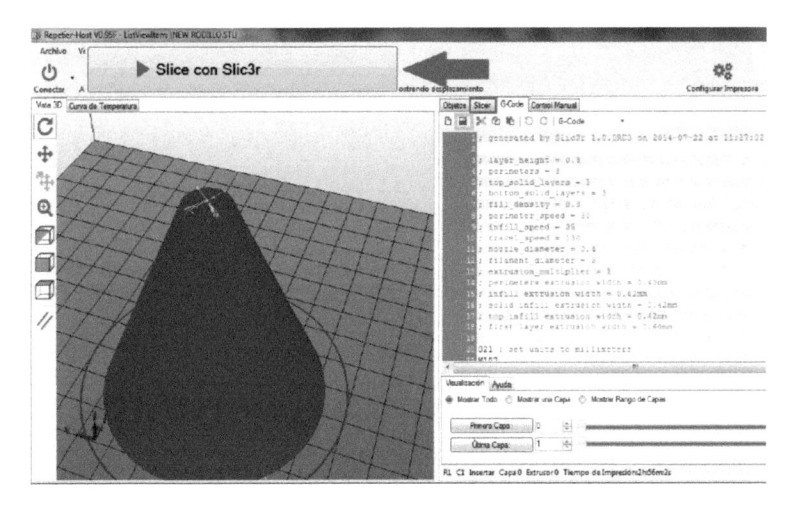

Fig. 9 3D printing simulation and G code generation

Table 1 Main printing parameter values

Printing parameters	Value
STL tolerance (mm)	0.007
Plastic material	ABS
Layer height (mm)	0.2
Bed temperature (°C)	100
Infill (%)	0.3

The input to Repetier software is a part or model in Stereolithography STL format. Repetier slices the object and generates the G-code to print a part according the user parameter definitions (Fig. 9). Table 1 shows the selected values used to print the oil mill components.

4.4 Resulting Oil Mill Mock Up

Figure 10 shows the assembled mock-up. A system of pulleys and an electric motor help to move the scaled oil mill. The reproduction is fixed to a flat support through suction pads that reduce vibration when the system is moving.

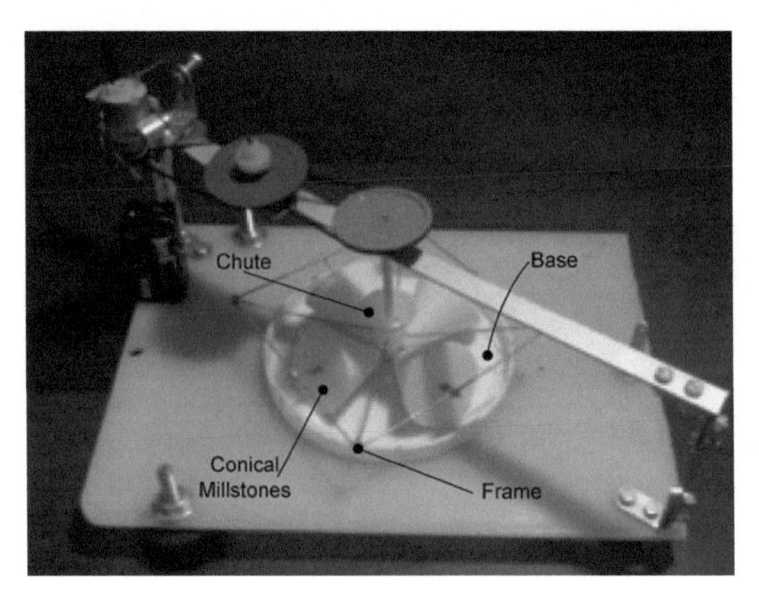

Fig. 10 Oil mill scaled mock-up

5 Conclusions

This work describes a methodology to obtain a scaled physical model of the grinding system of an eighteenth century oil mill.

Computer-Aided Design (CAD) programs help to model ancient machines. On the other hand, low-cost rapid prototyping methods as Fused Deposition Modelling (FDM) 3D printing provide physical reproductions of CAD models. Therefore, CAD and FDM can contribute to technical knowledge conservation.

We use Solidworks program to draw the grinder components (Figs. 4 and 10), and to assemble them. An FDM machine prints the chute, conical rollers and the grinder base. This machine runs a G code program provided by Repetier-Host software for each component.

Regarding to the frame that joins the printed "millstones", it is made of steel wire. An electric motor and a system of pulleys allow moving the mechanism. A set of suction pads fixes the mock-up to a flat surface.

References

1. Laroche, F., Bernard, A., Cotte, M.: Methodology for simulating ancient technical systems. Int. Rev. Numer. Eng. Integr. Des. Prod. **2**(1–2), 9–28 (2006)
2. Sánchiz, J.M.C.: Arqueólogos en la fábrica. Breve recorrido por la historiografía de la arqueología industrial, SPAL: Revista de prehistoria y arqueología de la Universidad de Sevilla **16**, 53–68 (2007)
3. Laroche, F., Bernard, A.: How to inject ancient know-how for future design: using advanced industrial archaeology during pedagogical projects. DPPI (2009)
4. Hudson, K.: Industrial Archaeology: An Introduction. Routledge Library Editions (2014)
5. Peters, C., Spring, A.P.: Digital Heritage, Industrial Memory and Memorialisation. Reanimating Industrial Spaces: Conducting Memory Work in Post-industrial Societies 212–234 (2014)
6. Laroche, F., Bernard, A., Cotte, M.: Advanced industrial archaeology: a new reverse-engineering process for contextualising and digitising ancient technical objects. Virtual Phys. Prototyping **3**(2), 105–122 (2008)
7. Laroche, F.: Advanced industrial archaeology and techno-museology: a new virtual life for industrial heritage. TICCIH Bull. **41** (2008)
8. Pujol, T., Solá, J., Montoro, L., Pelegrí, M.: Hydraulic performance of an ancient Spanish watermill. Renew. Energy **35**(2), 387–396 (2010)
9. Rojas-Sola, J.I., López-García, R.: Engineering graphics and watermills: ancient technology in Spain. Renew. Energy **32**(12), 2019–2033 (2007)
10. Tejero-Manzanares, J., Garrido-Sáenz, I., Pérez-Calle, M.D., Montes-Tubío, F.: La reconstrucción virtual en la recuperación del patrimonio metalúrgico de minas de Almadén. DYNA-Ingeniería e Industria **88**(3) (2013)
11. Dimitrov, D., Schreve, K., De Beer, N.: Advances in three dimensional printing-state of the art and future perspectives. J. New Gener. Sci. **4**(1), 21 (2006)
12. Villalba, F., Becerra, F., Expósito, E.: Aspectos socioeconómicos del aceite de oliva en Andalucía. El aceite de oliva virgen: Tesoro de Andalucía. Servicio de publicaciones de la Fundación Unicaja, Málaga, 79–105 (2009)
13. Kapellakis, I.E., Tsagarakis, K.P., Crowther, J.C.: Olive oil history, production and by-product management. Rev. Environ. Sci. Bio/Technol. **7**(1), 1–26 (2008)
14. Gibson, I., Rosen, D.W., Stucker, B.: Additive Manufacturing Technologies: Rapid Prototyping to Direct Digital Manufacturing. Springer Science & Business Media (2009)
15. http://www.reprap.org

An Introduction to the Ancient Mechanical Wind-Instrument Automata

Yu-Hsun Chen, Jian-Liang Lin and Hong-Sen Yan

Abstract Mechanical music-playing automata have been made to play music or simulate animal sounds for several thousand years, the combined products of music and technology. This study introduces the musical elements and mechanical design parameters of such objects. Three types of wind-instrument automata are classified based on the sounds they make, namely the whistle, pan flute, and slide whistle type. A structural analysis of the mechanisms of four typical ancient mechanical devices is also provided.

Keywords History of machines · Music-playing automaton · Mechanical aerophone · Mechanism structural analysis

1 Introduction

In the historical development of mechanical devices, many objects have been made that can simulate animal motions or sounds, and such devices are called automata. An automaton is a mechanical device capable of acting, such as carrying out certain motions or producing certain sounds, without any external motive force or operation. While these devices have no practical use in industrial manufacturing, agricultural work and daily life, they represent sophisticated mechanical designs and forms of artistic expression.

Musical automata are one well-known kind of automata, and have been produced for thousands of years. Although many examples have not survived, we known about them based on records regarding their structures and design theories in ancient literature. Figure 1a is an example of one such lost automaton, known as Singing Birds and Owl, designed by Hero of Alexandria (10-70 AD) [1]. There are also a number of automata that have been perfectly preserved, such as the

Y.-H. Chen · J.-L. Lin · H.-S. Yan (✉)
Department of Mechanical Engineering, National Cheng Kung University 1, University Road, Tainan 70101, Taiwan
e-mail: hsyan@mail.ncku.edu.tw

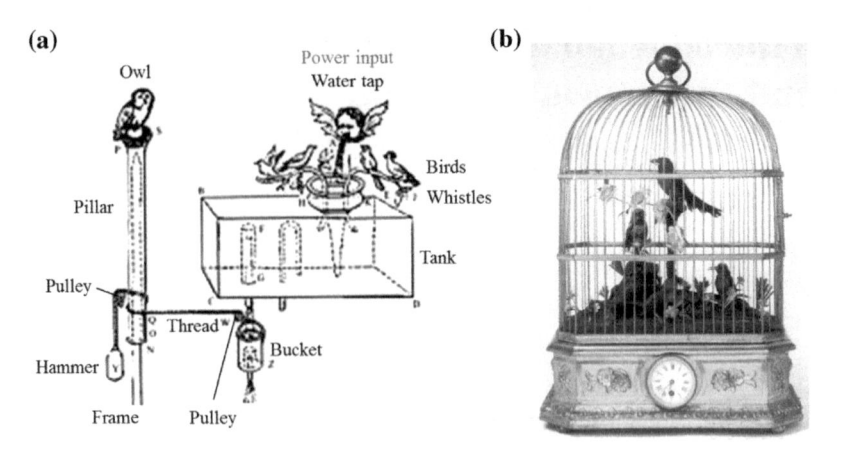

Fig. 1 Ancient Ancient musical automata. **a** Singing birds and owl [1]. **b** Bird cage clock [2]

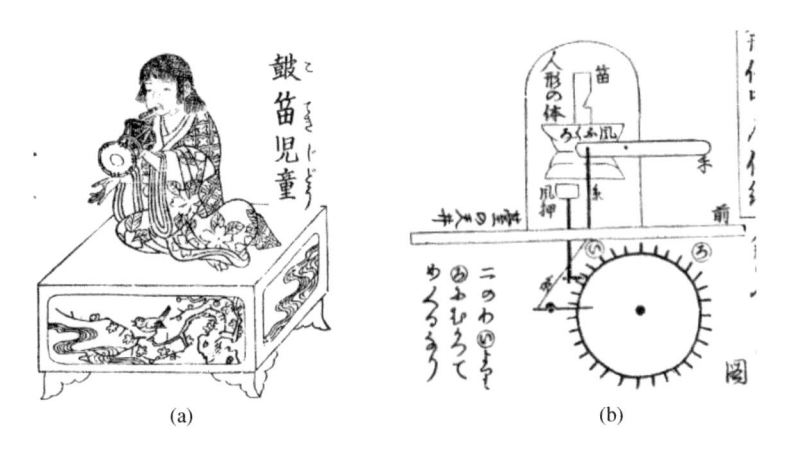

Fig. 2 Japanese automaton with wind-instrument [3]. **a** Exterior. **b** Mechanical structure

Wood-and-Plaster Bird Cage Clock in the Beijing Palace Museum, as shown in Fig. 1b [2]. In addition, Fig. 2 shows a flute-playing figure designed by Hosokawa (Japan, 1741-1796 AD) [3], while the Arabian author Al-Jazari (1136-1206 AD) also sketched designs for an automaton [4].

The designs of musical automata depend on the development of musical instruments, and there are many different ways of classifying these. Such methods consider the physical properties of the instrument (e.g., material, color, and shape), the means by which the music is produced with the instrument (e.g., striking, blowing or plucking), and its place in an orchestra or other ensemble. Based on an ancient system named "Natya Shastra," dating from between 200 BC and AD 200, musical instruments can be classified into the following four types: stringed instruments (chordophones), skin-head percussion instruments (membranophones), wind instruments (aerophones), and non-skin percussion instruments (autophones) [5–7]. These various types of musical instruments have also been used in music-playing automata.

This paper focuses on the study of wind-instrument automata. The fundamental elements of musical instruments are investigated, especially for aerophones. The elements that are critical to the design of wind-instrument automata are then considered. Finally, details of the classification and structural analysis of several famous ancient wind-instrument automata are presented.

2 Fundamental Elements of Aerophones

Musical sounds consist of the following four elements: pitch, duration, intensity, and timbre. These depend on the sound wave, and correspond to different physical characteristics. Pitch is related to frequency, duration is related to time, intensity is related to amplitude, and timbre is related to waveform. Table 1 shows the design parameters of wind-instrument automata and the related musical elements.

Pitch refers to the different sounds used to construct a melody, and a special unit system is used in musical theory to measure this. In general, pitch can be understood by frequency. Frequency (as measured in Hz) is related to the geometric size of musical instruments. More accurately, the sounding theory of musical instruments is based on the sizes of the resonant objects, and this concept can be applied to almost all types of musical instruments. For example, aerophones produce sound by the resonance of air in the tubes of the instrument, and different lengths of tubes can produce different pitches. As such, musicians have to press the holes or keys on the musical instruments so as to achieve a particular effective length of air tube to produce the resonance needed to the corresponding pitch. This concept was naturally applied to the design of wind-instrument automaton.

Duration (as measured in seconds) is the time period of a continuous sound. From a physical viewpoint, it is related to the natural decay period of a resonance, if there is no external force to restrain or stop this. However, the duration is also determined by the composition, as well as the ability of the musician. Therefore, with regard to aerophone musical automata the duration is regarded as being based on the process of movements the device makes. This is achieved by the mechanical controllers in ancient wind-instrument automata.

Table 1 Design parameters for wind-instrument automata

Music elements	Design parameters
Pitch	Effective length of whistle
Duration	Timing arrangement of controller
Intensity	Air flow rate
Timbre	N/A

Intensity (as measured in dB) is the loudness of sounds, and in ancient times there was no appropriate unit for this, with intensity being a comparative concept. From physical viewpoint, the intensity is the amplitude of a sound wave, and it is determined by exciting external force. For musical instruments, the intensity depends on the flow rate of air, the percussive or the plucking force, and even the capacity of the resonance box, with such concepts being applied to the design of musical automata. For automatic wind instruments, the control of intensity is achieved by the use of sophisticated mechanical designs that are a combination of an air-blower device and a specific mouth shape to control the air flow.

There is no unit for the quality of sound, or timbre, and different musical instruments have different sound qualities or wave forms. It can thus be said that timbre is determined by materials and geometry of musical instruments, and the same is also true for musical automata.

As noted above, the pitch, duration and intensity are variable elements that can be controlled by designing the movements of the various mechanisms with musical automata. In contrast, the quality of sound, or the timbre, is decided by the materials used, and is not related to the designs of the musical mechanical devices. As a result, the analysis of the mechanisms used in musical automata that is presented in this work will focus on the aspects of pitch, duration and intensity, and not timbre.

3 Classification and Structural Analysis

A wind-instrument automaton consists of an air-blower device, a set of whistles or flutes, and a control system. Based on the sounds they make, these automata can be classified into three types, as shown in Fig. 3, with examples of each type also provided below.

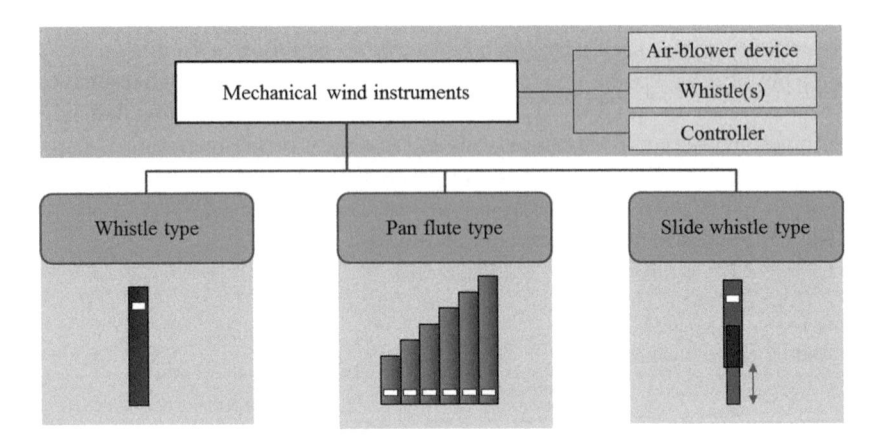

Fig. 3 Structural analysis and classification

4 Whistle Type

A mechanical device that makes a monotone sound is called a whistle type automaton. The Singing Birds and Owl device designed by Hero of Alexandria in the 1st century AD is an example of this, as shown in Fig. 1a. This wind-instrument automaton is powered by water and makes use of a water tap, a tank, and two whistles with birds. When water flows into the tank, the air that is originally contained in the tank is pushed out through the whistles under the birds and makes chirping sounds. On the other hand, the sounds stop when the tank is filled with water. Although there are two whistles attached to this device, they cannot make sounds at separate times. In addition, the duration and intensity of sounds depend on the water flow rate. The larger the flow rate, the faster the tank is filled and the shorter the duration. Moreover, when the flow rate of air passing through the whistles becomes greater, then this makes the sounds louder.

The second example is an automaton called Mechanical Shrine with a Singing and Revolving Bird, which was also designed by Hero in the 1st century, as shown in Fig. 4a. This wind-instrument automaton consists of a frame (link 1, K_F), a pulley with a wheel (line 2, K_U), a thread (line 3, K_T), and a slider with a whistle (link 4, K_P). The thread hangs the hemispherical slider and wraps around the pulley. When a person turns the wheel, the pulley rotates, the slider falls down and pushes out the air under the slider through the whistle to make a chirping sound. Figure 4b shows a structural sketch of the device. The duration of sound depends on the slider stroke, and the intensity is related to the falling velocity of the slider.

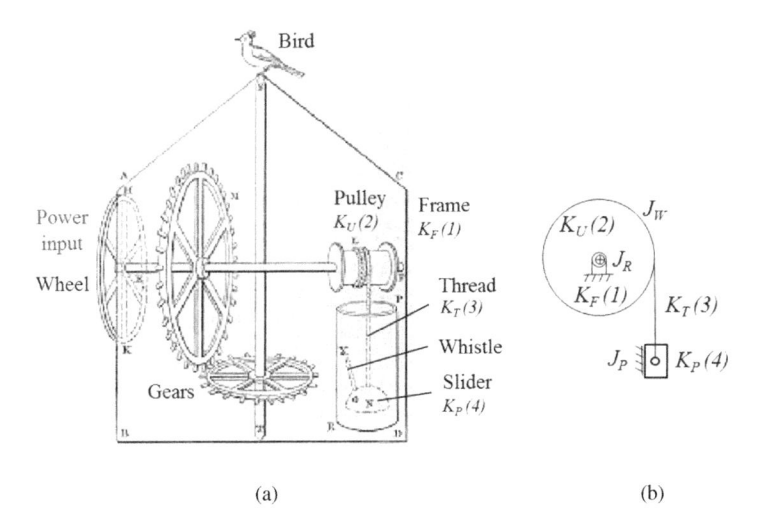

Fig. 4 Mechanical shrine with a singing and revolving bird. **a** Original illustration [1]. **b** Structural sketch

Fig. 5 Diagram of mechanical pan flute

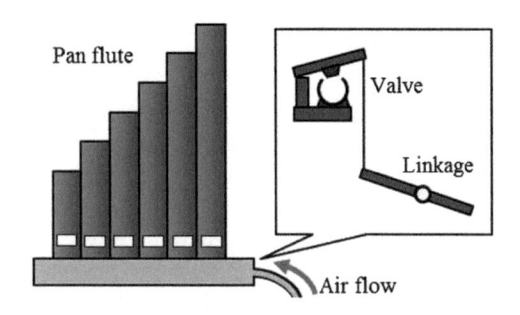

5 Pan Flute Type

A pan flute is made of many flutes with different effective lengths, thus producing multiple tones. A pan flute type mechanical instrument is thus able to automatically play a melody. This type of automaton is generally controlled by a cylinder with pins or fillisters, and the arrangement of these is programmable. Figure 5 shows a diagram of a mechanical pan flute. For each whistle, a valve is set at the end of the tube and connects to a linkage. The air flow can go through the whistle only when the valve opens.

The Cuckoo Bird and Pan Flute Player, invented by Athanasius Kircher (1602-1680 AD), is an example of such a device, as shown in Fig. 6a. This automaton is driven by water, and the air-blower device consists of a water tap, a tank, and an air tube. When water flows into the tank, the air contained in it is pushed into the tube at the upper end and this then supplies the air flow to the pan flutes. Water also flows through the tank and pours on to the water wheel to make it turn. Although there are three sets of pan flutes in this automaton, the mechanisms of each whistle in the pan flutes are the same. These include a frame (link 1, K_F), a pinion (line 2, K_G), a cylinder with fillisters (line 3, K_A), a follower (link 4, K_{L1}), a thread (link 5, K_T), and a linkage (link 6, K_{L2}). Figure 6b shows the structural sketch of the automaton. When the water wheel and the pinion turn, the gear and cylinder rotate in opposite directions, and the follower oscillates along the fillisters. The sound output is based on oscillation of the linkage connected to the valve. Therefore, if the cylinder rotates to the position that makes a follower go into a fillister, the corresponding valve opens and makes the whistle sound. Consequently, the duration of each sound depends on the length of the fillister, and the intensity is related to the water flow rate.

6 Slide Whistle Type

A slide whistle type mechanical instrument is a device that plays multiple tones using only one whistle, and the effective length of this whistle can be changed by a piston inside it. The size of mechanical slide whistle instruments are thus much

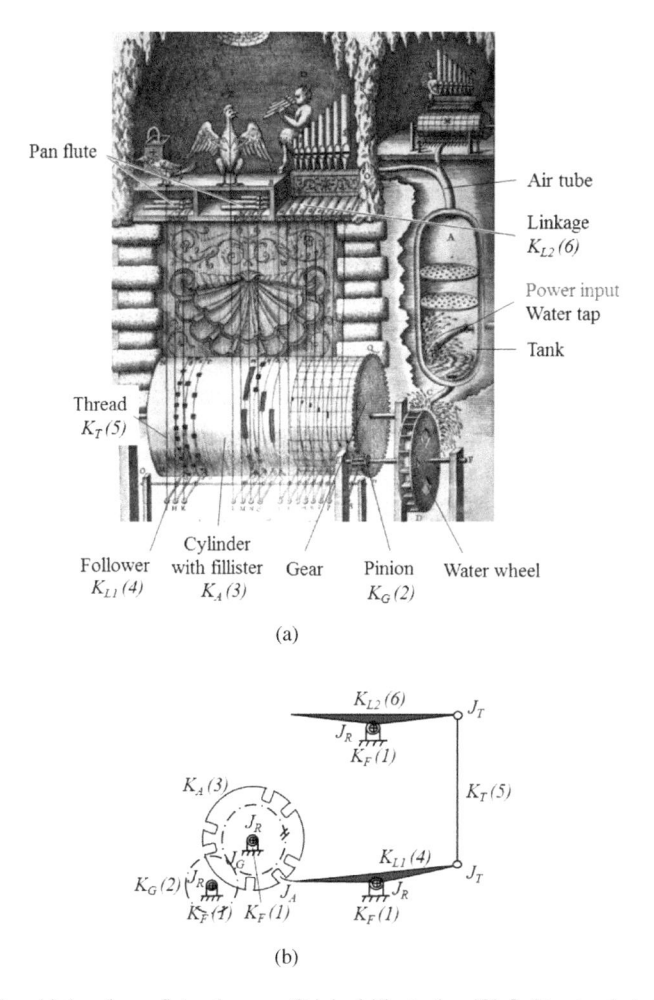

Fig. 6 Cuckoo bird and pan flute player. **a** Original illustration [8]. **b** Structural sketch

smaller than that of the pan flute type. Figure 1b shows a clock with a singing bird made in the 18th century. This wind-instrument automaton is driven by a coiled spring and includes two mechanisms, namely the controller and the air-blower [2, 9].

The structure of the controller mechanism is shown in Fig. 7a, including a frame (link *1*, K_F), a cam (line *2*, K_A), a follower (line *3*, K_{L1}), a linkage (line *4*, K_{L2}), and a piston inside the whistle (link *5*, K_P). When the cam is driven, the follower oscillates with the profile of the cam, and makes the piston slide through the transmission of the mechanism. Therefore, the effective length of the whistle depends on the piston's position. A structural sketch of this device is shown in Fig. 7b. The pitch of sounds is controlled by this cam mechanism to simulate the sounds of bird chirping.

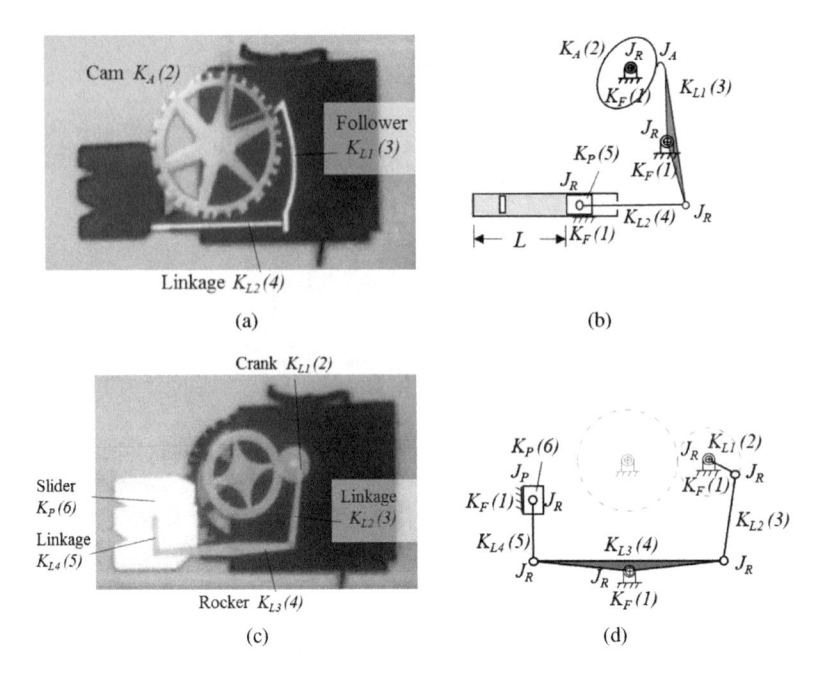

Fig. 7 Wood-and-plaster bird cage clock. **a** Diagram of controller [9]. **b** Structural sketch of controller. **c** Diagram of blower [9]. **d** Structural sketch of blower

The mechanism of an air-blower device is shown in Fig. 7c, and this includes a frame (link *1*, K_F), a crank (link *2*, K_{L1}), two linkages (link *3*, K_{L2} and link *5*, K_{L4}), a rocker (link *4*, K_{L3}), and a slider (link *6*, K_P). When the crank rotates, the rocker oscillates, and the slider reciprocates to pump the air. A structural sketch of the air-blower mechanism is shown in Fig. 7d. Both the duration and intensity of sound depend on the slider stroke.

7 Conclusions

A mechanical music-playing device is an automaton for performing music or producing the sound of birds chirping. For wind-instrument automata it is necessary to make air flow through a whistle to make sounds. The geometric size of the air tube, the controller used to carry out the movement procedure and the power sources used to move the device are closely related to the pitch, intensity and duration of each sound, and can be manipulated so as to produce a melody. The designs of these elements are thus the most ingenious parts of wind-instrument automata.

The fundamental elements of music are introduced in this paper, and the relationships between these elements and the mechanical parameters of automata are discussed. The pitch, duration, and intensity are mainly related to the effective length of the whistle, the timing arrangement of the controller, and the air flow rate, respectively.

Based on the sounds made by the wind-instrument automata, three types are classified in this study, namely whistle, pan flute, and slide whistle. In addition, structural analysis of four example automata is also provided in this paper. These automata are composed of an air-blower device, whistle(s), and controller, and many different mechanical components are used, such as pulleys, gears, cams, and linkages. To sum up, with regard to the pitch of sound the whistle type automata produce a monotone, while the pan flute and slide whistle automata produce multiple tones. The sound intensity and duration are influenced by the air-blower device, but they can also be related to the controller. However, the whistle and slide whistle types produce chirping sounds, and only the pan flute type of mechanical instruments can play melodies, because of their programmable cams. These automata combine music and machines, and reflect the level of technological development in the periods when they were produced.

Acknowledgments The authors are grateful to the Ministry of Science and Technology (Taipei, Taiwan) for its financial support of this work under Grant MOST 104-2221-E-006-007-.

References

1. Hero of Alexandria: Woodcroft B (translated). Create Space Independent Publishing Platform, London, The Pneumatics of Hero of Alexandria (2009)
2. Lang, X.H., Qin, X.P.: Clocks and Watches of the Qing Dynasty—From the Collection in the Forbidden City. Foreign Languages Press, Beijing (2002)
3. 細川半蔵, 菊池俊彦(annotated), 機巧図彙(in Japanese), 恒和出版, Tokyo, 166–169 (1796/1976)
4. Ibn Al-Razzaz Al-Jazari, Donald R. Hill (translated), The books of knowledge of ingenious mechanical devices, Elite Publishers Ltd., Karachi (1049/1988)
5. Montagu, J.: Origins and Development of Musical Instruments. The Scarecrow Press, Metuchen (2007)
6. Kartomi, Margaret: J, On Concepts and Classifications of Musical Instruments. University of Chicago Press, Chicago (1990)
7. Rault, L.: Musical Instruments: A Worldwide Survey of Traditional Music-making. Thames & Hudson Ltd, London (2000)
8. Kircher, A.: Musurgia Universalis, Rome (1650)
9. Introduction of the preserved clocks and watches, demonstration video, Beijing Palace Museum, July 15, 2014

Dynamic Reconstruction of a Colonial Mexican Mechanism

J.C. Jauregui-Correa and G. Rodriguez-Zahar

Abstract In this paper, a detail analysis of a Colonial mechanism is presented. In this work, a procedure for reconstructing the dynamic parameters of antique mechanisms is described. With this analysis, the restoration of antique artifacts incorporates actual loading conditions based on engineering data. The reconstruction of the engineering data included field measurements, X ray analysis, functional analysis and interpretation of the kinematic chains. The mechanism that is analysed in this work was originally design by miners during the Colonial time. The procedure presented here can be applied to other mining machinery, it can reconstruct the engineering concepts that were used before the industrial revolution.

Keywords Antique mechanisms · Reconstruction procedure · Kinematics · Identification techniques

1 Introduction

Before the independence of Latin America (The beginning of XIX), the economy was based on mining, agriculture, textile, commerce and other craftsman activities [1, 2]. From this period, mining machines were the most developed ones in the region; unfortunately, there are few records that support the engineering knowledge [3]. In general, the mechanisms were built with wood reinforced with iron linkages and were power with water mills or mules. Within the mining areas, other mechanism were also developed, some of these mechanism were used in religious activities and the ingenious mining craftsmen constructed the "Christ of the three falls". This mechanism is used during the Easter procession and it represents this

---

J.C. Jauregui-Correa (✉)
Universidad Autonoma de Queretaro, Queretaro, Mexico
e-mail: jc.jauregui@uaq.mx

G. Rodriguez-Zahar
Instituto Nacional de Antropologia e Historia, Guanajuato, Mexico
e-mail: gzahar@mac.com

episode in some cities in Mexico, Spain and other countries. To make the procession more attractive, there is a mechanism under the platform that moves the statue and simulates that Christ falls. This type of mechanism were first identified around 1596 in Mexico City [4]. Other researchers have identified this type of mechanisms in different parts of Mexico, most of them related to mining towns [5].

The reconstruction of engineering data from ancient machinery is a topic of interest in many countries. One of the oldest sophisticated mechanism is the Antikethera mechanism [6]. A detail analysis of this mechanism is presented by Yan and Lin, they applied their method for reconstructing the synthesis of the interior mechanism of the Antikythera [7]. They identified the specific types of members and joints, and transformed the generalized kinematic chain into a specialized chain. Koetsier described different autonomous mechanisms that were developed from the ancient Greece until the XIX century. He only presents a description of the mechanisms without presenting a kinematic or dynamic analysis [8]. Penta et al. presented a mechanical study of the "carroballista" used during the Roman Empire. They utilized modern analysis tools to determine original design parameters. The engineering analysis confirms much of the historical data about the use of this weapon [9]. Ceccarelli described the most significant mechanism that were developed during the Renaissance in Italy [10]. Hsiao presents a method for the synthesis of an antique crossbow. He classified three type of joints in ancient mechanism: similar joints as modern machinery, joints that are uncertain and joints that are rare in modern mechanisms [11]. Chen et al. presented another analysis of ancient Chinese mechanism, the wooden horse carriage, and they applied Yan's approach for the synthesis of lost ancient Chinese mechanisms [12]. Thomas presented the analysis of Leonardo Torres' endless mechanism [13]. Wauer et al. described the work done by Ferdinand Redtenbacher, who was one of the pioneers in formalizing the kinematic analysis of mechanisms [14].

Since it was almost impossible to locate documents that describe the mechanism presented in this paper, and archeological reconstruction of existing samples were conducted. Reconstructing the engineering data requires conventional procedures and sophisticated treatments. Azema et al. presented method for analyzing the welding process used by Romans to produce large bronze statues. They used ultrasonic phased arrays to study antique statues [15]. The problems of recovering historical pieces are incrustations, corrosion and post-corrosion after exposing them to the open air. Veprek et al. presented a method for restoring corroded archeological artifacts. They showed results on iron, silver and bronze [16].

Figure 1 shows the actual statue while it is in a procession. The original designs are still in operation in some areas of Mexico. The entire mechanism consist of a Christ representation (a moving statue), the cross and the Cyrene that also holds the cross. The mechanical parameters were obtained by physical measurements and x ray analysis. The next section describes the archaeological procedure for gathering the data and the following sections describe the mechanical synthesis and the dynamic model.

Fig. 1 Picture of the procession

2 Procedure for the Reconstruction

The need for engineering data arise from the restoration process of an antique statue and its mechanism. The entire mechanism consisted of a moving plate that rotates the statue, a drum with a cable that helps controlling the velocity and a roller, at the Cyrene hand, for sliding the cross. It was necessary to reconstruct all the design parameters and the functionality of the existing artifact. Figure 2 shows the

Fig. 2 Procedure for reconstructing the dynamic model

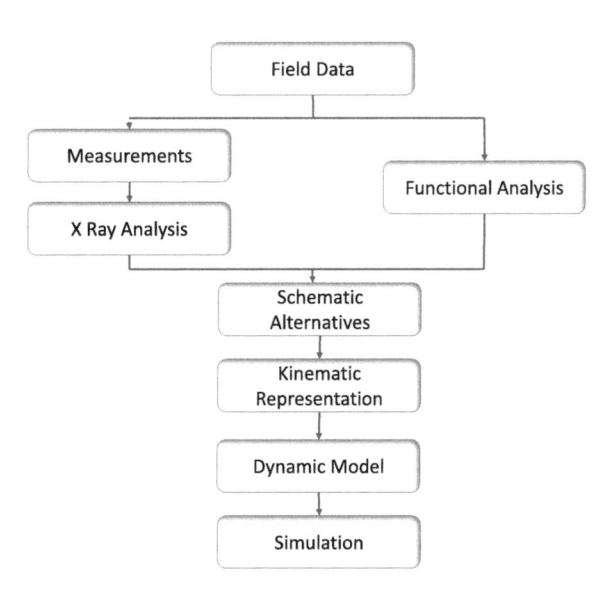

procedure for recovering the design parameters and technical data. The first stage of the process consisted on measuring the actual dimensions from both, the statue and the acting platform. Since the statue can be morph to different postures, it was necessary to explore its internal structure. The internal structure was analyzed using X ray images. Once the dimensions and internal structures were obtained, a schematic representation was drawn. Based on this scheme, a kinematic model was determined, it was assumed that the statue only moves in one plane. As mention by Hisao [6], some of the joints were determined assuming certain behavior. One of the most difficult part of the statue is the waist flexibility, because it has a metallic support that acts as a spring and the final position is adjusted with ropes. The other joints were easily represented as rotating pairs and prismatic pairs. A kinematic model was derived using the geometric method and the dynamic model was obtained from the kinematics. The external loading was estimated based on the functional analysis, the mechanism is operated manually; thus the loading was estimated from the human carrying capacity.

The dynamic model was simulated using Matlab®.

3 Field Measurements and Functional Analysis

3.1 *Functional Analysis*

Figure 3 shows a picture of the statue before it is configure for the representation of the "three falls". From this figure, it is possible to see two hinges and the ropes that limits the waist rotation. Although the arms are free to move, during this procession they are attached to the cross and the main body.

Fig. 3 Statue before the configuration for "Christ of the three falls"

Fig. 4 The statue dress as Christ, carrying the cross and Cyrene holding the back of the cross

Once the statue is "stiffened", it is mounted on the moving platform. The moving platform is carried on shoulders, and it has hinge that rotates the entire statue, simulating the Christ falls. Figure 4 shows the final stage of mounting the statue on the platform. In this picture, it is possible to see the connecting point where the cross is attached to the statue, and the roller that enables the cross to slide (hand of the Cyrene).

Figure 5 shows the hinge that holds the statue's foot. The hinge is lowered manually and it has two chains that limits the movement. In order to return the statue to its original position, a drum and a cable roll up the statue. The detail of the drum is shown in Fig. 6. The returning mechanism (drum and cables) are attached to statue as shown in the same figure. This is an auxiliary mechanism that helps controlling the speed of rotation of the statue.

Fig. 5 Hinge that rotates the entire statue

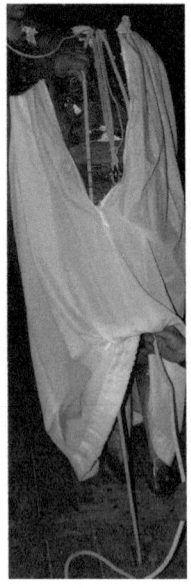

Fig. 6 Detail of the drum and control cables

3.2 Internal Analysis

In order to determine the flexibility of the waist, an X ray analysis was conducted. Figure 7 shows the picture of the X ray. From this figure, it was possible to see that the waist has a spring effect and it is limited with the cables shown in Fig. 3. Once the cables are tensioned, the statue losses its flexibility and it is assumed that it moves as a rigid body.

Fig. 7 X ray analysis of the waist

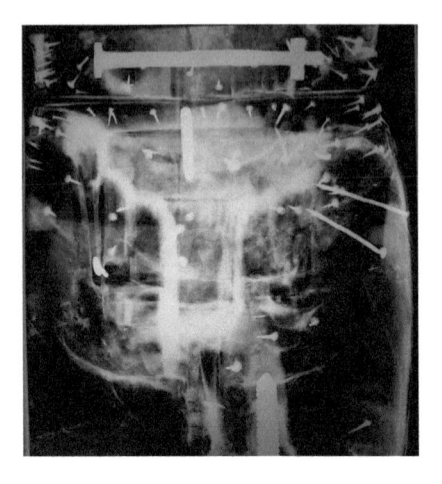

4 Kinematic and Dynamic Models

From previous measurements and analysis, it was possible to represent the mechanism in a sketch. Figure 8 shows two sketches, one in the upright position and the other in the bent position.

Based on these schemes, the kinematic model was constructed as described in Fig. 9. The mechanism is divided into two mechanism, each one associated with the control systems. The first one is a RRPR mechanism, and the joints are: the hinge, the attachment between the cross and the statue, and sliding roller at the Cyrene hand. The second mechanism is also a RRPR mechanism, it also includes the hinge at the platform, the attachment between the cross and the statue, the cable (considered as a prismatic joint) and the drum.

The kinematic and dynamic models are determined using the notation of Fig. 10. The dimensions were taken from the statue and the supporting platform (Table 1).

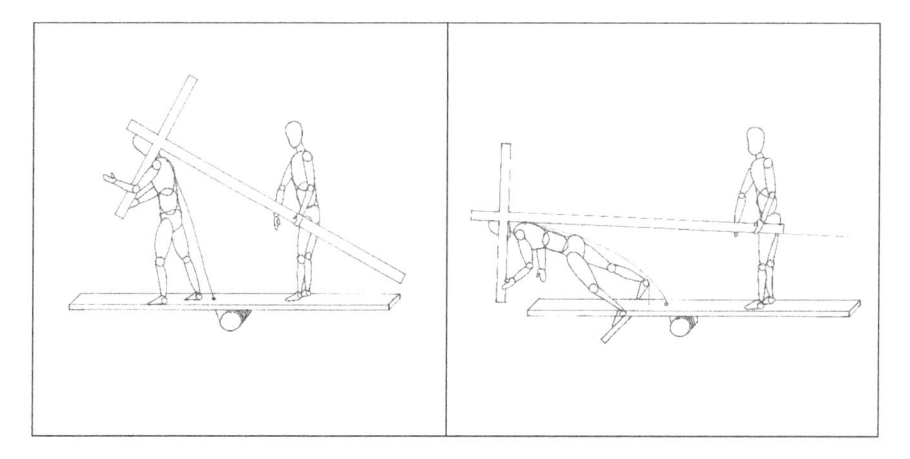

Fig. 8 Schematic representation of the extreme positions

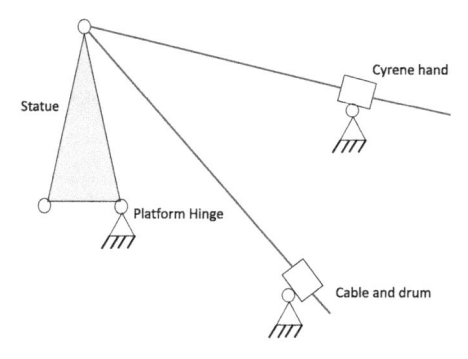

Fig. 9 Kinematic model

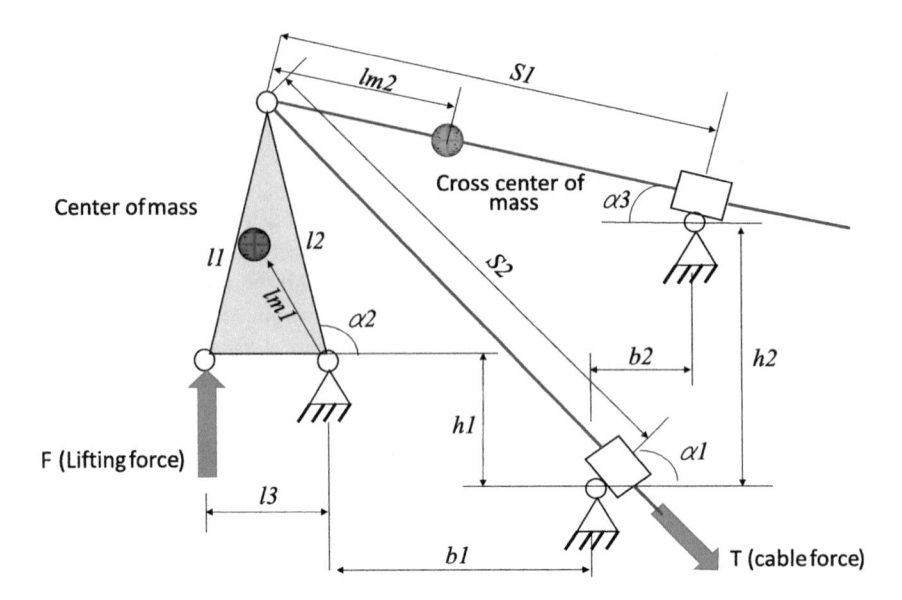

Fig. 10 Dynamic model

Table 1 Dimensions

Term	Dimension (mm)
l1	1450
l2	1450
l3	900
lm1	1150
lm2	400
b1	250
b2	950
h1	250
h2	1300

In this model, the variables are: the length of the cable ($S2$), the distance from the shoulder to the Cyrene hand ($S1$), the angular position the cable ($\alpha 1$) and the angular position of the cross ($\alpha 3$). Since the movement of the statue is controlled from the platform, the independent variable is $\alpha 2$.

4.1 Kinematic Model

The kinematic model is found using the well-known geometric method. Although the joints are very loose and have a large backlash, for this analysis, it is assumed that there is no backlash. Therefore, from Fig. 10, the length of the cable is:

$$S_2 = \sqrt{l_1^2 + b_1^2 + h_1^2 + 2l_1(h_1 sin(\alpha_2) - b_1 cos(\alpha_2))} \qquad (1)$$

and its angular position is:

$$tg(\alpha_1) = \frac{l_1 sin(\alpha_2) + h_1}{l_1 cos(\alpha_2) - b_1} \qquad (2)$$

The velocity and acceleration can be derive from these equations. They are omitted due to space limitations.

The position of the statue is controlled by two persons, one moves the platform and the other ties the cable with the drum (Figs. 5 and 6). Since the angular acceleration is difficult to determine, it is assumed that the platform rotates at constant speed. Therefore, the acceleration of the center of mass is:

$$\bar{a}_1 = lm_1 \dot{\alpha}_2^2(-cos(\alpha_2)i - sin(\alpha_2)j) \qquad (3)$$

The cross has a more complex movement. The position of the cross is found using the notation of Fig. 10. The distance between the shoulder (attaching point) and the Cyrene hand is calculated as:

$$S_1 = \sqrt{S_2^2 + b_2^2 + h_2^2 - 2S_2 h_2 sin(\alpha_1) - S_2 b_2 cos(\alpha_1)} \qquad (4)$$

and the angular position is:

$$tg(\alpha_3) = \frac{S_s sin(\alpha_1) - h_2}{S_2 cos(\alpha_1) - b_2} \qquad (5)$$

In this case, the acceleration of the center of mass is calculated as;

$$\bar{a}_2 = \left[-\ddot{S}_1 cos(\alpha_3) + 2\dot{S}_1 \dot{\alpha}_3 sin(\alpha_3) - (lm_2 - S_1)\alpha_1^2 cos\alpha_3\right]i \\ + \left[\ddot{S}_1 sin(\alpha_3) + 2\dot{S}_1 \dot{\alpha}_3 cos(\alpha_3) + (lm_2 - S_1)\alpha_1^2 sin\alpha_3\right]j \qquad (6)$$

where i and j are the unit vectors along the horizontal and vertical directions.

4.2 Dynamic Model

The dynamic model is determined using the Newton-Euler method. For the statue, the equilibrium equations are:

For the statue, the force in the horizontal direction are:

$$R_{01x} + R_{21x} = m_{c1}\mathbf{a}_{1x} - T_{cab}\cos(\alpha_1) \tag{7}$$

and in the vertical direction:

$$R_{01y} + R_{21y} = m_{c1}(\mathbf{a}_{1y} + g) - F_{LF} - T_{cab}\sin(\alpha_1) \tag{8}$$

The moments around the hinge are:

$$\begin{aligned}R_{21x}l_1\cos(\alpha_2 - \pi/2) &+ R_{21y}l_1\sin(\alpha_2 - \pi/2) = m_{c1}(\mathbf{a}_{1y}+g)l_{m1}\sin(\alpha_2 - \pi/2)\\&+ m_{c1}\mathbf{a}_{1x}l_{m1}\cos(\alpha_2 - \pi/2) + T_{cab}\sin(\alpha_1)l_1\sin(\alpha_2 - \pi/2)\\&- T_{cab}\cos(\alpha_1)l_1\cos(\alpha_2 - \pi/2) - F_{LF}l_3\end{aligned} \tag{9}$$

For the cross, the force in the horizontal direction are:

$$R_{02x} + R_{21x} = m_{c2}\mathbf{a}_{2x} \tag{10}$$

and in the vertical direction:

$$R_{02y} + R_{21y} = m_{c2}(\mathbf{a}_{2y} + g) \tag{11}$$

The moments around the Cyrene hand are:

$$\begin{aligned}R_{21x}S_1\sin(\alpha_3) + R_{21y}S_1\cos(\alpha_3) &= m_{c1}(\mathbf{a}_{1y}+g)(S_1 - l_{m2})\cos(\alpha_3) + m_{c1}\mathbf{a}_{1x}(S_1\\&- l_{m2})\sin(\alpha_3) + m_{c1}\mathbf{a}_{1x}(S_1 - l_{m2})^2\ddot{\alpha}_3\end{aligned} \tag{12}$$

In this way, a system of six equations and six unknowns is obtained. The unknowns are the reactions at the hinge (vector $\mathbf{R}_{01}$), the Cyrene's hand (vector $\mathbf{R}_{02}$) and the connection at the statue's shoulder (vector $\mathbf{R}_{21}$). These equations are useful for finding the maximum forces as a function of the platform rotating angle (α_2). The values of the lifting force and the tension on the cable were determined considering the force that an adult can apply at that position (Fig. 10). In this case it was assumed that

$$\begin{aligned}F_{LF} &= 150\text{ N}\\T_{cab} &= 200\text{ N}\end{aligned} \tag{13}$$

The model was solved using Matlab® assuming that $90° < \alpha_2 < 135°$.

5 Results

Only the acceleration of both centers of mass are presented. Figure 11 shows the statue's horizontal and vertical accelerations. The vertical component increases more than the horizontal component. Figure 12 shows the acceleration of the other center of mass. In this case, a similar pattern is found.

The dynamic model was solved from the kinematic results (Eqs. 7–13). Figure 13 shows the variation of the force at the connecting point. The connecting point is where the cross is attached to the statue. In this case, the maximum force is 624 N and it occurs at $\alpha_2 = 135°$. Figure 14 shows the variation of the force at the hinge. This is the most loaded point in the mechanism. And the lowest forces occur at the Cyrene's hand (Fig. 15).

From a practical point of view, it is important to determine the force at the connecting points. In this study, it was important to know the forces applied at the shoulder in order to select the best materials for restoring the statue.

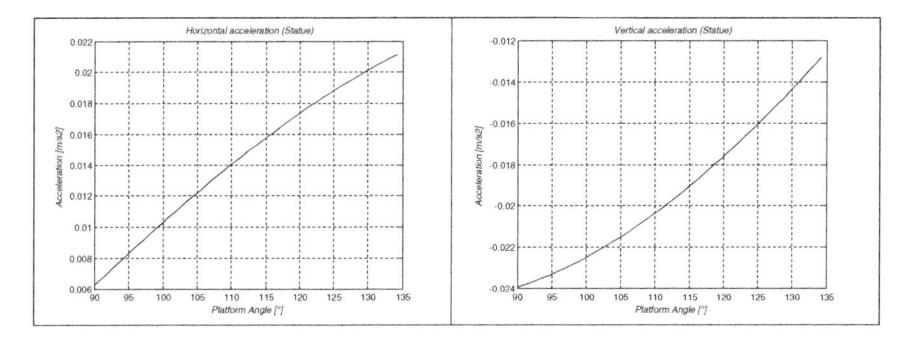

Fig. 11 Acceleration of the statue's center of mass

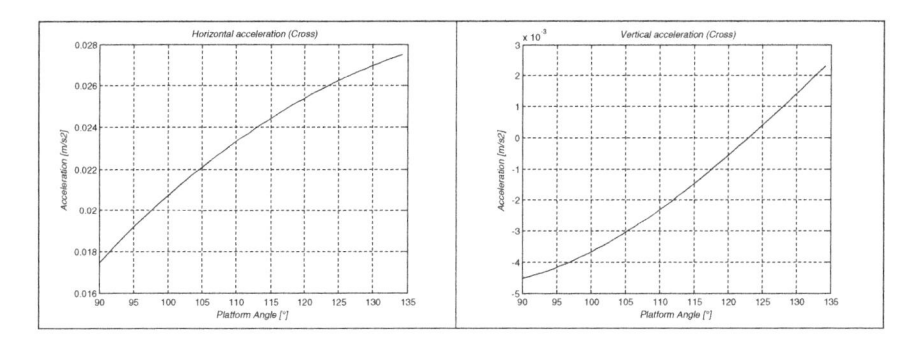

Fig. 12 Acceleration of the cross's center of mass

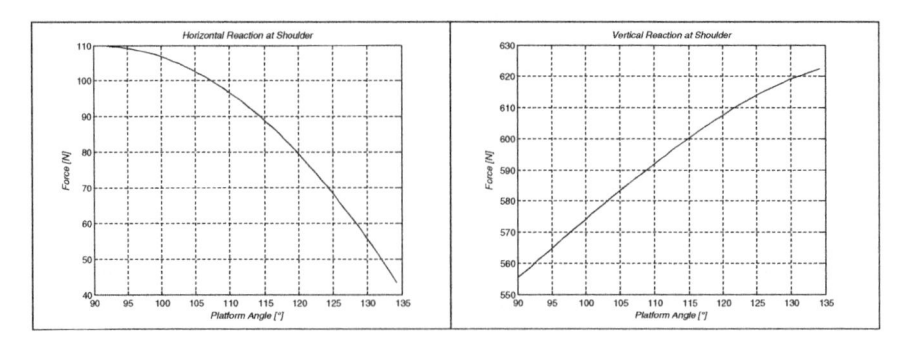

Fig. 13 Dynamic forces at the shoulder's attachment

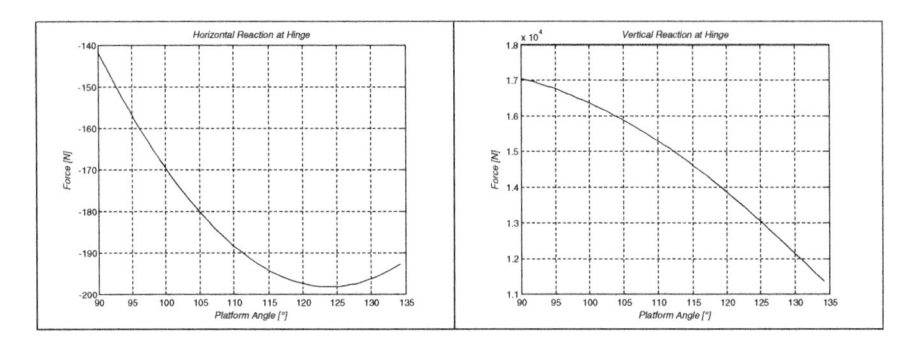

Fig. 14 Dynamic forces at the platform's hinge

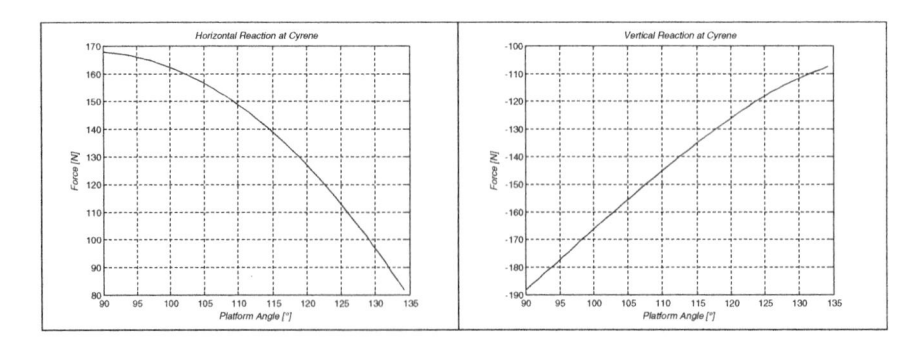

Fig. 15 Dynamic forces at Cyrene's hand

6 Conclusions

Recovering engineering data from antique mechanisms requires a detail kinematic and dynamic analysis. Unfortunately, most of this mechanism were built without keeping records of their dimensions and functionality. In order to recover this information, it is necessary to use modern concepts to identify the kinematic functions of this mechanism. In many cases, it is possible to represent the antique mechanism with the kinematic chain theories, the big limitation is the actual shape of the linkages, the unknown flexibility of the linkages, backlash and the out of plane movements.

In this paper, a detail analysis of a Colonial mechanism is presented. It was possible to reproduce the design concepts used before the independence of Latin America, and it was possible to reconstruct the dynamic behavior of this mechanism. With this analysis, the restoration of antique artifacts incorporates more accurate loading conditions.

The procedure presented here can be applied to other mechanism, particularly the reconstruction of engineering data from mining machinery could be used to understand the social knowledge before the independence.

References

1. Baewell, P.J.: Silver Mining and Society in Colonial Mexico, Zacatezas 1546–1700. Cambridge University Press, Cambridge (1971)
2. Garcia, C.L.: Tecnología Herramental y Maquinaria Utilizadas en la Producción Monetaria Durante el Virreinato. Investigaciones Sociales, pp. 93–121 (1998)
3. Flores, R.S.: Historia de la Tecnología y la Invensión en México. Fomento Cultural Banamex, Mexico (1980)
4. Padilla, A.D.: Historia de la Fundacion y Discurso de la Porvincia de Santiago de Mexico de la Orden de Predicadores. Academia Literaria, Mexico (1955)
5. Jasso, A.E.: Imágenes en Caña de Maiz, San Luis Potosi: Universidad Autonoma de San Luis Potosi (1996)
6. Moussas, X.: Antikythera Mechanism The oldest computer and Mechanical Cosmos 2nd Century BC, 1st edn. University of Birmingham, Birmingham (2014)
7. Yan, H.-S., Lin, J.-L.: Reconstruction synthesis of the lost interior mechanism for the solar anomaly motion of the Antikythera mechanism. Mech. Mach. Theory **70**, 354–371 (2013)
8. Koetsier, T.: On the prehistory of programmable machines: musical automata, looms, calculators. Mech. Mach. Theory **36**(5), 589–603 (2001)
9. Penta, F., Rossi, C., Savino, S.: Mechanical behavior of the imperial carroballista. Mech. Mach. Theory **80**, 142–150 (2014)
10. Ceccarelli, M.: Renaissance of machines in Italy: from Brunelleschi to Galilei through Francesco di Giorgio and Leonardo. Mech. Mach. Theory **43**(12), 1530–1542 (2008)
11. Hsiao, K.: Structural synthesis of ancient Chinese original crossbow. **37**(2), 259–271 (2013)
12. Chen, F.C., Tzeng, Y.F., Chen, W.R., Yan, H.S.: On the motion of a reconstructed ancient Chinese wooden horse carriage. Mech. Mach. Theory **58**, 165–178 (2012)
13. Thomas, F.: A short account on Leonardo Torres' endless spindle. Mech. Mach. Theory **43**(8), 1055–1063 (2008)

14. Wauer, J., Moon, F.C., Mauersberger, K.: Ferdinand Redtenbacher (1809–1863): pioneer in scientific machine engineering. Mech. Mach. Theory **44**(9), 1607–1626 (2009)
15. Azéma, A., Angelini, F., Mille, B., Framezelle, G., Chauveau, D.: Ultrasonic phased array contribution to the knowledge of the flow fusion welding process used to make the Roman large bronze statues. Weld. World **57**(4), 477–486 (2013)
16. Eckmann, C., Elmer, J.Th., Veprek, S.: Recent progress in the restoration of archeological metallic artifacts by means of low-pressure plasma treatment. Plasma Chem. Plasma Process. **189/190**(4), 221–466 (1988)

Leibniz's Developments of Machine Science

A.R.E. Oliveira

Abstract This paper is a tribute to the tercentenary of Leibniz's (1646–1716) death. After Aristotle (384–322 BC), one of the most impressive and original accomplishments of building rational knowledge about nature can be found in Leibniz's dynamics. Furthermore, his science of motion occupies a prominent place in his thought. In addition to these theoretical developments, Leibniz designed a calculating machine which performed the four arithmetic operations and worked as a mining engineer in Harz (Germany), with the design and construction of windmills. In this paper Leibniz's contributions to machine science, namely dynamics and machine design, are presented and discussed.

Keywords History of scientific revolution · The science of motion · History of dynamics · History of machines

1 Introduction

Gottfried Wilhelm Leibniz [1] was born at Leipzig on July I, 1646, the son of Friedrich Leibniz, Professor of the University of Leipzig. His mother, Catharina Schmuck, daughter of a Professor of Law, as your father both came from a noble and academic family. Since the age of seven he was allowed to read his father's library containing books of poets, orators, historians, jurists, philosophers, mathematicians and theologians.

During his lifetime, Leibniz produced books and treatises of great value on the widest possible range of subjects. This is a characteristic of his work.

Leibniz's work remained unknown for a long time. Only at the beginning of the twentieth century would some of his important books and documents finally appear. Practically at the same time two fundamental books were published. Bertrand Russell's [2] *A Critical Exposition of the Philosophy of Leibniz,* which appeared in

A.R.E. Oliveira (✉)
Polytechnic School of Rio de Janeiro, Rio de Janeiro, Brazil
e-mail: agamenon.oliveira@globo.com

London in 1900, and Louis Couturat's [3] work *Logique de Leibniz*, published in Paris in 1901. Other important books include E. Cassirer, *Leibniz' System*, published in Marbourg in 1902 as well as the Gueroult [4] book, *Dynamique et Métaphysique Leibniziènne*, published in Strasbourg in 1934.

The seventeenth century was a period of rapid social and scientific development in which the Aristotelian science that had dominated the Middle Ages and Renaissance schools was replaced by classical physics [5]. To understand this transformation from an Aristotelian framework to a new vision of the physical world, it is fundamental to study the role that Leibniz played. According to Aristotelian physics the basic explanatory principles are matter and form. Together these two principles compose the idea of body. However, according to Leibniz's doctrine [6] *force and body are directed solidly against the mechanist doctrines of Descartes, Hobbes and their followers. There is every reason to think that it was largely in place by 1686 when, seemly out of nowhere, Newton published his* Philosophiae naturalis principia mathematica [7], *which came to be known simply as the* Principia, *and developed a conception of force that was very different from what Leibniz had developed. The ultimate success of the Newtonian program has all but driven Leibniz's concept of force off the playing field.*[1]

One of the central interests of Leibniz's philosophy was the understanding of physical world. He was one of the most important physicists of the late seventeenth century. Apart from Newton (1642–1727) there is no other physicist of his generation who contributed more to mathematical physics. Thus, to understand the history of sciences in this crucial period, we must understand Leibniz's theories of physics [8].

2 Leibniz the Reformer of Rationalism

Leibniz commenced his investigations with a profound knowledge of mathematics, law, scholastics, as well as a special preference for logic and theological meditation which he did not separate from science. To the contrary, he tried to unify science and Christian faith. In this context it is important to emphasize the question of the existence of God, the most central problem of Catholic theology, for Leibniz and Descartes. The central contrast between the two philosophers is the degree of theocentrism in their respective philosophies. Leibniz's theocentrism was absolute, while Descartes [9] was only theocentric as far as his philosophy of Self was concerned. Both philosophers, with their encyclopedic spirits, used scientific and empirical methods to prove the existence of God, notwithstanding their rationalism.

Leibniz reacted against Locke's (1632–1704) empiricism and reinforced Cartesian rationalism. However, he recognized the importance of experience for

---

[1]Garber [6] 99–179.

knowledge, but what was essential to him was the reasoning power which organized thought.

His passion for logic was expressed in his desire to understand everything. Everything could be understood. He believed that there was perfect agreement between thought and reality. As a consequence the method to be followed in order to investigate reality was logical deduction and geometrical reasoning. The logic of necessity and according to him another logic, the logic of the probable, were the same as the logic of the truth in the moral sciences and in history. All this reason reasoning about logic and methodology led him to a kind of hierarchy in science. Initially it was necessary to study the nature of man and medicine. Second, the history of mankind, and finally the technics of arts, the connection between theory and application. Poetry did not appear in his considerations.

Nature for him was the first objective of study. The world could not be explained by a fatal or arbitrary mechanism. It was submitted to a need which surpassed logic and geometry, this need being of a metaphysical order: it was the result of choosing wisdom.

In his critique of Bacon (1561–1626), Descartes [10], and the mechanists in general, in his *Principles of the Nature and Grace*, he reestablished the idea of final causes. Neither being nor the world were reduced, as in Descartes' doctrine, to an extension, rather they were force, energy, mind, perception and life. From inanimate things to animal species, on to man, nature produced a gradual and continuous effort to achieve consciousness. But the world is not only matter and motion. It is mainly energy which belongs to it and is conserved. In fact, in his vision these forces are spiritual forces which are in 'pre-established harmony,' according to the will of God, and build the best of the possible worlds.

3 Leibniz and the Scientific Revolution

Seventeenth century thought is characterized by the reaction against Aristotle, not the Aristotle of antiquity, but Aristotle as seen through the eyes of the medieval scholastics. Medieval philosophy is part of theology. With few exceptions of Avicena (980–1037) and Averroes (1126–1198), medieval thinkers did not consider themselves philosophers. For them, the philosophers were the ancient pagan writers like Plato and Aristotle. Curiously, their theology used the methods and logical categories and techniques from the old philosophers which implied serious theological questions within the doctrine [11].

In this reaction there appeared the mark of what has come to be called empiricism, the attitude of the new science which appeared in the Renaissance and a characteristic which persists to the present among a certain school of thinkers. However, not every reaction against medievalism was of an empirical nature. Descartes' (1596–1650) philosophy [12], for instance, was a typical product of the age and a direct result of this fight against scholasticism. Indeed, in some aspects it was the direct antithesis of empiricism.

The seventeenth century Scientific Revolution [13] is a theme that continues to attract the attention of many historians of science. These portray this process as occurring roughly in the following sequence: Copernicus' (1473–1543) reformulation of Ptolemy's (100–170) solution of the problem of planets with the need to restore their lost harmony; Kepler (1571–1630) and Galileo's (1564–1642) acceptance of its realistic proposition; based on this perspective the development of mathematical tools to study the heavens; the mathematization of free fall and projectile motion to confirm the realistic basis of Copernicanism [14]; and the development of a new inertial conception of motion, associating an abstract idealized concept of nature, linked to empirical and artificial means of experiment.[2]

This revolution was mainly a revolution in scientific method which took place during the Renaissance culminating with Newton's *Principia* [15–17]. It had a double aspect: (1) first, scholastic notions of essence and final causes were abandoned and a new explanation of phenomena was sought in efficient causation and the assessment of quantitative change. Changes were to be explained solely in terms of matter and motion and the mechanical character of natural phenomena was expressed in quantitative terms. Measurement was the instrument of scientific discovery [18]. (2) Second, it became clear that the discovery of efficient causes and the assessment of quantitative change could be made only by means of observation and experiment and not any more by argument from first principles. This was the only fruitful method for the discovery of the structure of the natural world and the interaction of its parts, the details of which could not be explained by any general principles.

Descartes' metaphysics sets out the first of these two assumptions of Renaissance science, while Locke's (1632–1704) epistemology is the theoretical expression of the second presupposition.

In relation to the assessment of quantitative change, as is well known, the progress in infinitesimal calculus developed independently by Newton and Leibniz during the Scientific Revolution was a fundamental part of this and was crucial for the transformations towards a new physics [19, 20]. The capacity to solve old and new problems as well as to create modern new branches of knowledge, also demonstrated the power of the scientific method.

With respect to the development of infinitesimal calculus, there emerged in 1682 the great scientific review *Acta Eruditorum*, in which Leibniz would publish the results of his scientific investigations of the foundation of this new branch of mathematics. 1682 saw the first publication of his work in *Acta Eruditorum*. In 1684 what is considered his first work on differential calculus was published: *A new method for maximum, minimum and tangents*. In this paper Leibniz used for the first time the symbol of differential d and listed the rules for the differentiation of addition, subtraction, product and quotient, the chain rule, the second differentiation, the method of separation of variables for solving differential equations.

---

[2]Cohen [13] 21–147.

Furthermore, Leibniz abandoned the expression *methodus tangentium directa* and adopted the term *differential calculus* [21].

4 The New Leibnizian Conceptual Framework

Leibniz made his first scientific synthesis in 1671, which would be radically reviewed when he also carried out what he called his *reformatio* of mechanics [22]. At the beginning of 1676, towards the end of his time in Paris, this was done in his *De corporum concursu*, when Leibniz tried to examine the problem of the laws of motion in depth, especially collisions between bodies, in an attempt to move beyond the relativistic point of view expressed by Huygens (1629–1695) and Mariotte (1620–1684) with respect to theoretical mechanics based on geometry.

Leibniz believed that he knew the key to reconciling empirical laws and the a priori principle of the quantity of motion conservation similar to Descartes' work.

This key was the principle of equivalence between total cause and complete effect. Leibniz then intended to combine both equivalents in terms of both definitions of cause and effect which would permit a single uniform measure of driven force for all possible cases to be obtained. He believed that this measure would be the product of mass by velocity [23].

In this sense *De corporum concursu* is a unique work from the epistemological point of view. Leibniz then adopts a systematic 'deduction' of the laws of collision in order to establish conformity with the Cartesian principle of quantity of motion conservation; he compared the theoretical calculations with the deduction of results from an experience based on pendulum properties to measure their displacements effects under the collision of unequal masses.

Initially Leibniz reformulated mechanics using a hypothetical principle of conservation, which considered that there is a fall in height equivalent to the product of the mass multiplied by the velocities of the bodies previously in collision. He also believed that if the explanation was constructed according to the equivalence postulation between total cause and complete effect, as well as the methodological rule of continuity, the theory would work without a proper conceptualization of the underlying forces. This aspect guided Leibniz to reform mechanics by developing a theoretical system of concepts and arguments which maximized the coherence and functionality of this new mechanics [24]. However, some people only see it as a subtle differentiation between concepts on the frontiers of science and metaphysics.

According to Duchesneau,[3] when Leibniz produced his first global formulation of metaphysics, this represented the official birth of reformed mechanics. This can be seen in some articles in his *Brevis demonstratio erroris memorabilis Cartesii*

---

[3]François Duchesneau is Professor of philosophy at Montreal University. He published several important books regarding Leibniz's scientific and philosophical works: *Leibniz et la méthode de la science*, PUF, Paris, 1993; *La dynamique de Leibniz*, Librairie Philosophique J. Vrin, Paris, 1994.

et aliorum circa legem naturalem (1686), while the same arguments appear in *Discourse of metaphysics*, written in the same year. The context is given by concerns with the laws of nature.

Brevis demonstratio is characterized by the opposition of Leibniz's reformed dynamics and Cartesian science to its argumentative structure. The turning point towards dynamics appears in *Phoranomous* (1689), where Leibniz applies the equivalence between total cause and complete effect. He is also mainly concerned with a dynamic science explained by action and force. This dynamic structuration organized as a true science is completely revealed in his *Dynamica de potentia* (1689–1690). In this great demonstrative synthesis, definitions and heuristic principles allow the conception of multiple theoretical constructions with the capacity to take into account empirical laws. Thus, Leibniz follows an a priori way to demonstrate the theorem of the conservation of driven action.

In 1695, *Specimen dynamicum* revealed another aspect of the Leibnizian method, the theoretical construction of his dynamics. In appearance this text follows an a posteriori path for demonstrations. In addition the typology of primitive and derivative forces allows the integration of formal components of force: *conatus*, impetus, *vis viva*.

In *Essay de dynamique tardive* (1700) we can see the last phase of the theoretical structuration of Leibniz's dynamics. He shows the integration of the relative conservation principles in the absolute principles of conservation. The synthesis of these models is based on architectural principles.

During 1690–1700 and 1700–1710 the question of a system of a priori proofs was emphasized within Leibnizian dynamics.

5 Principle of Vis Viva Conservation

If we look at the reformation of dynamics, developed by Leibniz in January 1678, we are mainly concerned with how we can characterize his use of the expression mv^2, which Michel Fichant[4] designates as "the nucleus of a complete doctrinal set" and its consequences for metaphysics. After this the text *De corporum concursu* written in January 1678 gives a definitive answer to Leibnizian dynamics.

Eight years before the *Brevis demonstratio erroris mirabilis Cartesii* and the *Discourse of metaphysics,* Leibniz arrived at his canonic definition of force, as well as the formulation of his conservation principle. As we know, his main motivation was to establish the rules governing the collision of two bodies which implied the substitution of the product mv by mv^2 as a measurement of force.

Nonetheless, according to Professor Michel Fichant (1941), it is possible to describe the logical propositional structure underlying the process of *vis viva* construction as follows: if the combination of the rule of translation of the center of gravity with the constancy of relative velocity is true, then, it is not (always) true that $\sum m|v|$ is constant. Although the identification of the quantity effect which

Fig. 1 Leibniz's demonstration of the vis viva principle

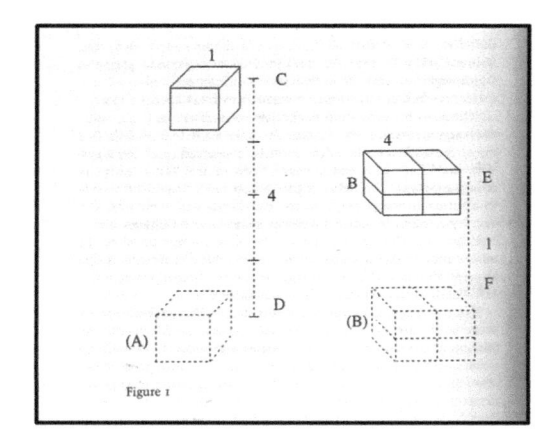

measures force by means of the height of the lifting or falling thereby provides a new formula for the forces of conservation according to mv^2. Distance conservation, or the relative velocity after collision, provides the second formula as a starting point according to which the theory of elastic direct collision can be developed.

Leibniz uses several other arguments to contradict the Cartesian quantity of motion that is conserved. Using for instance the principle of the equality of cause and effect, he postulates that if the quantity of motion is conserved, one could build a perpetual motion machine, a machine that would create the ability to do work out of nothing at all. Leibniz is concerned not only to show that quantity of motion differs from force but, that quantity of motion is not conserved. This, it is the quantity mv^2 that correctly measures force and mv^2 is conserved in the world.

Using Fig. 1, Leibniz argue that if two bodies A and B have different quantities of motion, their size multiplied by the square of their speeds will be equal. It is easy to generalize this and show whenever they have equal force, the size multiplied by the square of their speed will be equal and that whenever this is violated, the ability to do work will either be gained or lost, violating the principle of the equality of cause and effect. (See S. D., part I, par. 16, G. M. VI 244-45: A. G. 128).

A more complete discussion of this problem can be found in the article *Leibniz: physics and philosophy* in [1].

5.1 Application of the Vis Viva Concept to Machines

One application of the concept of living forces (*vis viva* or kinetic energy) was an important step towards the development of the machine science. According to Navier (1785–1836), the first study where we find the principle of the conservation of living forces applied to machines was *Hydrodynamics* [25], published by Daniel Bernoulli (1700–1782) in 1738. This important achievement was ignored by physicians and engineers for several decades. Only after the memoir of Claude

Borda (1733–1799) entitled *Memoir on the Hydraulic Wheels* appeared in 1767, did the principle of the conservation of living forces start to be applied to machines. He was the first to apply this principle to hydraulic wheels. Some years later, in 1781, Coulomb published his *Theoretical and Experimental Considerations on the Effect of Windmills* using the same principle [26].

According to Navier, Borda and Coulomb's contributions were fundamental steps and showing remarkable progress in regard to Bernoulli's *Hydrodynamics*. Navier also emphasizes that there was a need for the creation of a general theory involving that principle with the capacity to calculate machine efficiency. It is exactly in this context that he affirms that this theory was created by Lazare Carnot (1753–1823) in his *Fundamental Principles of Equilibrium and Motion*, in 1803 [27]. He also attributes to Carnot the general demonstration of the theorem which calculates the loss of living forces due to collisions between non-elastic bodies (hard bodies).

Carnot then studied the problem of the transformation of work in motion by considering all the parameters involved. From this viewpoint, this meant established convenient variations of the terms of the quantity FVT, i.e., the moment of activity, later denominated in Coriolis' work[4] (1792–1843). If time is the most important parameter and it should be minimized, the effect must be produced in a very short time. It is possible to generalize this reasoning for the case of a system of forces, for instance; if we have the forces F, F', F'' with the velocities V, V', V'', acting during the times T, T', T'', respectively, then one reads:

$$FVT = F'V'T' = F''V''T'' = PH$$

If the motion of each one of the forces is variable, we will take the quantity: $\int (FVdt + F'V'dt' + F''V''dt'')''$, or, if we have the forces directions with respect to velocities, one has:

$$\int [FVdt \cos(F \wedge V) + F'V'dt' \cos(F' \wedge V') + F''V''dt'' \cos(F'' \wedge V'')]$$

This is the definition of work done by all forces [28].

The quantity PH, the effect to be produced by a machine, is, by Carnot called latent living force. Obviously, it presupposes a transformation of living forces from latent to actual one. If we call M the mass of the weight P, and V the velocity correspondent to a height H, one reads:

$$PH = 1/2MV^2$$

The relation above is always valid for any variation of the effect. When Carnot presented the above equation he mentions Leibniz as being its author and says that

---

[4]Coriolis [29] 1–34.

only after Leibniz were the forces acting in bodies in motion calculated in a different form than the equilibrium situation.

After these achievements, which made an important contribution to the development and application of the conservation of living forces principle, Navier made some remarks about and additions to Belidor's (1698–1761) *Hydraulic Architecture*. He demonstrated the principle of the conservation of living forces in a single mass but also generalized this result to a system of n particles using the d'Alembert principle.

However, a great evolution in the living forces principle occurred with the publication of Coriolis' book *Du Calcul de l'Effet des Machines* in 1829 [29]. This work is considered one of the most important nineteenth century works in mechanical engineering. The term *work* was coined and the constant ½ was incorporated in the expression of living forces in this book. A great advance in machine science was achieved with this book which did not consider the machine as a conservative system. The balance of living forces throughout the entire system was equal to global work also spent or produced by the system in the expression of living forces in this book. A great advance in machine science was achieved with this book which did not consider the machine as a conservative system. The balance of living forces throughout the entire system was equal to global work also spent or produced by the system.

6 Leibniz and Machine Design

6.1 *Windmill Construction*

In the 1679–1686 period, Leibniz worked in the silver mines of Harz, around one hundred kilometers from Hannover [30]. His mechanical projects were aimed at solving three problems related to mining services: ventilation, water elevation, and the transportation of materials from galleries. Leibniz began the association with the House of Hannover in 1676 where he was employed as Privy Councilor and librarian of the important ducal library. At that time John Frederick was duke of Brunswick-Lüneburg and ruled over the Principality of Calenberg, part of the duchy, from 1665 until his death in 1679.

The problem of finding a regular low cost energy source to take water out of mines drew at that time the attention of several engineers and scientists. In a letter to Denis Papin (1647–1712) in 1705, Leibniz enumerated five possibilities for doing this task: human work, animal force (mainly horses), falling water, wind, and fire. In this context he was interested in using wind as a driven force. Due to the changes in intensity and direction which characterize wind flow, he proposed to transform kinetic wind energy into potential water energy, so that an irregular accumulation of energy could become compatible with regular consumption. Obviously, the term energy was not yet used in the technical literature. As is well known the concept of energy only appeared in the middle of the nineteenth century.

Sometimes the name of Denis Papin appears associated to Leibniz's investigations on steam machines. This occurs because Papin worked on steam engines for several years. In 1695 he moved from Marburg to Kassel. In 1705 he developed a second steam engine with the help of Leibniz based on an invention of Thomas Savery. By January 1705, Papin had received Leibniz's sketch of Savery's engine. In March, Papin wrote to Leibniz:

"I can assure you that, the more I go forward, the more I find reason to think highly of the invention which, in theory, may augment the powers of man to infinity; but in practice I believe I can say without exaggeration, that one man by this means will be able to do as much as 100 others can do without it" [31].

It is important to emphasize some aspects of the scientific and economic problems postulated by Leibniz. One of these questions is the correct understanding of energy which is closely related to his principle of the conservation of living forces. As we know, this principle is a kind of bridge concept between theoretical and applied mechanics because of its fruitful use in the study of machines. With respect to the economic questions behind his technological and scientific problems we can say that the period under consideration is part of the beginning of the Industrial Revolution with Leibniz playing a role as engineer and entrepreneur.

Probably he was the first to use mechanical analogies in economic studies. In this sense something else that is well known is the analogy between potential energy accumulated in a reservoir with capital accumulation. Another important question belonging to the history of science is the importance attributed by historians to his period in Harz in the genesis of Leibnizian dynamics, which is identified with the practical problems studied by him in his engineering projects.

Finally, the idea of using an irregular source of energy, the wind, finding from the engineering point of view a form of a regular consumption, is recognized as one of the first studies of automation and control.

6.2 Calculating Machines

In 1671, Leibniz designed a calculating machine called the Stepped Reckoner [32]. It was actually first built in 1673, based on Pascal's (1623–1662) ideas and did multiplication by repeated addition and division by repeated subtraction. Two prototypes were built (see Fig. 2); today only one survives in the National Library of Lower Saxony in Hanover, Germany.

The above machine has the following characteristics:

- Structure: Two attached parallel parts, an accumulator section to the rear and an input section to the front. There is also an indicator and a control to reset the machine.

Fig. 2 Leibniz's machine

- Dimensions and materials: 67 cm long, 27 cm wide and 17 cm high. Polished brass and steel mounted in a large oak case with the dimensions: 97 cm × 30 cm × 25 cm.
- Operations: Add and subtract an 8 digit number to/from a 16 digit result; multiply two 8 digit numbers to obtain a 16 digit number; divide a 16 digit number by an 8 digit divisor.

7 Conclusions

Our main purpose here was to show the great importance of Leibniz's work for machines sciences, with the starting point that dominant Newtonian ideas since the hegemony of Newtonian mechanics has left Leibniz in a secondary position as a founder of mechanics.

The intellectual context where Leibniz lived is described, as well as his participation in the seventeenth century Scientific Revolution. However, the basis for a true revolution in the scientific conceptual framework was laid by Leibniz with the development of differential and integral calculus and dynamics. It is worth emphasizing the establishment of the conservation of living forces principle postulating the expression mv^2 as the quantity that remains constant in motion. Not only did this concept have great importance in theoretical mechanics but it also marked the birth of applied mechanics, as shown in where Lazare Carnot's contribution to a new theory of machines appeared. The principle of conservation formulated by Leibniz was an important contribution to the principle of conservation of energy which emerged in the nineteenth century.

References

1. Jolley, N.: The Cambridge Companion to Leibniz. Cambridge University Press, New York (1995)
2. Russell, B.: The Philosophy of Leibniz. Redwood Book, London (1900)
3. Couturat, L.: La Logique de Leibniz. (1902)
4. Gueroult, M.: Leibniz, Dynamique et Métaphysique. Paris. Auber Éditions (1967)
5. Westfall, R.S.: A Construção da Ciência Moderna. Porto Editora (2001)
6. Garber, D.: Leibniz: Body, Substance, Matter. Oxford University Press (2009)
7. Blay, M.: Les "Principia" de Newton. Presses Universitaires de France, Paris (1995)
8. Duchesneau, F.: La Dynamique de Leibniz. Librairie Philosophique Vrin, Paris (1994)
9. Descartes, R.: Princípios da Filosofia. Paris. Edições 70 (1997)
10. de Buzon, F., Carraud, V.: Descartes et les "Principia" II: Corps et Mouvement. Paris. PUF (1994)
11. Russell, B.: História da Filosofia Ocidental. Companhia Editora Nacional, São Paulo (1967)
12. Descartes, R.: La Géometrie. Éditions Jacques Gabay, Paris (1991)
13. Cohen, H.F.: The Scientific Revolution: A Historiographical Inquiry. The University of Chicago Press, Chicago (1994)
14. Hall, A.R.: From Galileo to Newton. Dover Publications, Inc., New York (1982)
15. Cohen, I.B.: Introduction to Newton's Principia. Harvard University Press, USA (1978)
16. Cohen, I.B.: Revolution in Science. Harvard University Press, Boston (2001)
17. Cohen, I.B.: The Newtonian Revolution. Cambridge University Press, UK (1980)
18. Blay, M.: La Science du Mouvement: De Galilée à Lagrange (2007)
19. Leibniz, G.W.: Oeuvres Concernant le Calcul Infinitesimal. Translated from Latin to French by Jean Peyroux. Albert Blanchard, Paris (1694)
20. Leibniz, G.W.: Oeuvre Concernant la Physique. Translated from Latin to French by Jean Peyroux. Albert Blanchard, Paris (1694)
21. Leibniz, G.W.: La Naissance du Calcul Differentiel. Librairie Philosophique Vrin, Paris (1995)
22. Leibniz, G.W.: La Reforme de la Dynamique. Librairie Philosophique Vrin, Paris (1994)
23. Leibniz, G.W.: Escritos de Dinâmica. Madrid Editorial (1991)
24. Costabel, P.: Leibniz et la Dynamique en 1692. Librairie Philosophique Vrin, Paris (2013)
25. Bernoulli, D.: Hydrodynamics. Dover Publications, Inc., New York (1738)
26. Coulomb, C.A.: Théorie des Machines Simples. Librairie Scientifique et Technique Albert Blanchard, Paris (2002)
27. Gillispie, C.C.: Lazare Carnot et sa Contribuition á la Théorie de l'Infini Mathematique. Librairie Philosophique Vrin, Paris (1979)
28. Oliveira, A.R.O.: A History of the Work Concept: From Physics to Economics. Springer, Amsterdam (2013)
29. Coriolis, G.G.: Du Calcul de l'Effet des Machines. Carilian-Goeury Librairie, Paris (1829)
30. Elster, J.: Leibniz et la Formation de l'Esprit Capitaliste. AUBIER. Éditions Montaigne, Paris (1975)
31. EIR Science & Technology, vol. 23, No. 8, Feb 16 (1996)
32. Wikepedia. Stepped Reckoner

On the Mechanics of Live Nature in the Works of V.P. Goryachkin

V. Chinenova

Abstract The famous scientist, academician V.P. Gorjachkin (1868–1935), the founder of agricultural machinery industry in Russia, posed and developed a global problem "Human—Machine—Environment" in his works. A specific property of most agricultural machines, including modern ones, is the fact that a human operator is needed to control these machines. Starting from his very early works in agricultural mechanics and theory of agricultural machinery, Goryachkin explored working movements of a human operator in search for optimal forms of these movements, which would minimize the burden on arms and legs of an operator. When solving these issues, it was necessary to find appropriate relationships between operator's movements and the physiology of the human being. This paper takes up three Goryachkin's publications on the mechanics of live nature and his thoughts about application of the methods of classical mechanics in agricultural engineering and biology.

Keywords Methods of classical mechanics · Theory of mechanisms · Mechanical tractors · Live movers · Live engines

1 Introduction

Goryachkin's works mentioned in this article are descriptive in nature. They are interesting primarily from the point of view of philosophy of science. Modern aspects of biomechanics and other branches of science are not touched on here. This is an example from the history of science, which partly reflects the situation in science in the early 20th century as well as an approach of a great scientist to the problem of interrelations between nature and technology.

In the late 1927, Vasilii Prokhorovich Goryachkin presented the report "Mechanics of Animated Nature" [1] at the annual meeting of the Timiryazev

V. Chinenova (✉)
Moscow State University, Moscow, Russia
e-mail: v.chinenova@yandex.ru

Agricultural Academy. This presentation was not occasional. For three quarters of a century gone, many manuals on applied mechanics were issued in which the forces of human beings and animals were studied as a source of energy. To the turn of the 20th century, however, the concept of human beings and animals as a source of physical force was no longer topical due to the growth of power availability per worker. Goryachkin's research was devoted to an important part of the theory of mechanisms, which has been developed nowadays in the theory of the structure of manipulators and robots. This subject was also considered in the article "Mechanical Tractors and Live Movers" published in his Collected Works [2], and the work "Operation of live engines" [3]. Goryachkin's figures used in this article.

Goryackin's article [3] deals with the movement of humans and animals. Goryachkin notes that while mechanical engines are mostly used to rotate a working shaft with more or less constant impedance, live movers, which may provide tractive force, can adapt to a variety of works that differ not only in kind of motion but also in needed force, velocity and duration.

He analyzes the conditions and character of the most efficient work of live movers and estimates the working efficiency of humans, horses, oxen, cows, camels and dogs. His conclusion is that the best conditions for their work are 8 h of maximal effort, 8 h of maximal speed and 8 h of rest.

2 Principles of Mechanics and Biology

Goryackin noted that muscle movements were analogous to motions of parts of machines, while movements of joints fully corresponded to the theory of kinematic pairs. At the beginning of his report, he focused on common aspects of mechanics and biology: "Mechanics is a science that studies motion; we should consider the motion not only as some visible displacement but also in a wider sense, as a change of one value in respect to another. This definition includes any phenomenon, whatever it could be, because it is nothing but a change of one value in respect to another. This implies that fundamental principles of mechanics should be necessarily observed in animated nature. Thus, biology and mechanics should develop in parallel ways because the main ideas are of the same origin" [1].

Further Goryachkin gave some examples to confirm this statement. This idea is far from being new; L. Euler mentioned three composing parts of a machine as far back as in the end of the 18th century.

The first element (a source of energy) conveys its energy to the second one (a receiver of energy), which receives the energy and serves as a working body that, in turn, passes the energy to the last element (an accumulator of energy) to accumulate the energy. Thus, if a plough is added to a horse, soil is also needed to accumulate the horse's energy in the soil via the tool. Goryachkin gave other examples of these relations: locomobile—thresher—sheaves; radiative energy—plant—fruit; feed—animal—mechanical work or products; labor—ground—capital.

Using this method of studying phenomena of all kinds, as emphasized by Goryachkin, a general layout is to determine the efficiency of the energy source, the working receptivity of the transmitting body, and the value of energy stored in the accumulator to find out the relationships between these elements and to evaluate them.

When we mention the simplicity of a phenomenon, we involuntarily recollect the fundamental Hamiltonian principle of least action or Gaussian principle of least deviation, which was formulated by E. Mach in his *Mechanics* as follows: only those events occur which can happen under given conditions and in the simplest way possible, i.e., with the least deviation from freedom.

Thus, if a heavy point *a* passes a way *ab* in vertical free fall, the same point will pass a way *ac* when it is forced to move under different conditions, along a slope, the path *ac* being a projection of the free-fall path *ab*. This forced path *ac* differs from the free-fall path *ab* by vector *bc*, which is a perpendicular to the slope and it is shorter than any inclined vector *bc'* and *bc"*. In other words, if the point *a* moving along the slope passed the path *ac'* or *ac"*, which is shorter or longer than the path *ac*, the deviation from the free-fall would be *bc'* or *bc"*, more than *bc* (Fig. 1).

When we mention the symmetry of organisms, we involuntarily recollect the law of action and counteraction. Since biological phenomena occur as a result of the action of internal forces, their external appearances should be bilateral and they are obliged to be symmetric, although this symmetry may be mechanical rather than geometrical. Thus, roots can be considered as mechanically symmetric to the stem and leaves.

Goryachkin's reflections about the mechanics of nature are interesting because they convey the spirit of science and show his creative laboratory.

Further Goryachkin gives some examples from biology. These may seem naive but one should not forget that the article was written in the early 20th century. It is particularly interesting from the point of view of history of science, reflecting ideas of scientists of that time and their attempts aimed at the application of mathematical methods to phenomena of the surrounding world.

When we say about limited nature of phenomena, about death and immortality, these ideas can be also understood from the mechanical point of view, because energy, whatever it may be, penetrates into an organism through surface and spreads in volume. For instance, light, warmth, and radiative energy in general

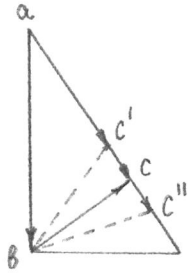

Fig. 1 A diagram illustrating the principle of least action

penetrate through outer surfaces of windows or organisms, nutrients are absorbed through the surface of the stomach, etc., spreading then in volume. For this reason, single-celled organisms are immortal because they proliferate by fission, the surface of a cell remaining unchanged.

Multi-celled organisms are bound to be mortal due to the following mathematical game of figures. Let us consider how the surface l^2 and volume l^3 change when changing the linear size l:

$$l = 0, 1/3, 1/2, 1, 2, 3, \ldots$$
$$\text{surface} \quad l^2 = 0, 1/9, 1/4, 1, 4, 9, \ldots$$
$$\text{volume} \quad l^3 = 0, 1/27, 1/8, 1, 8, 27, \ldots$$

This table shows that for l between 0 and 1, the square (l^2), that is, the surface is relatively larger than the volume (l^3); the surface becomes less than the volume after reaching the unity. Therefore, at the very beginning, a young organism has a rather large surface comparatively with a volume; the amount of energy that may penetrate into the organism at that time surpasses the amount that may be taken by the organism. Therefore, the process develops with acceleration when the linear size of the organism changes between 0 and 1. When the linear size reaches the unit and increases further, the surface inevitably lags behind the volume and the process decelerates.

Mathematically, accelerated development of a process is inevitably characterized by the concave curve OA and the decelerated development is characterized by the convex curve AB (Fig. 2). Thus, every phenomenon is characterized by the integrated curve OAB, on which the inflexion point A where one curve passes into another is a beginning of the organism's death. Therefore, the death is an inevitable consequence of internal, rather than external, reasons lying in the mathematical basis of processes.

If the development of different finite processes is represented in the form of an integrated curve, this allows us to make assertions and generalizations in many cases with full confidence. After establishing a form of the curve, we may think about composing an equation of the curve, not only a correlation relationship.

Fig. 2 A graph of an increasing and decreasing continuous function

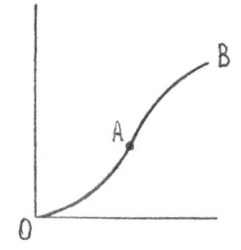

The form of the integrated curve, in and of itself, leads to further conclusions. For example, it may be proven that the so-called "gain", i.e., an increment in a unit of length or weight, has a simpler hyperbolic form, what predetermines results of many experiences.

When calculating the feed, as well as in many other cases, zootechnicians have already successfully used the principle of mechanical similitude.

The use of the ratio g/e, the weight of a person to its stature, in anthropometric measurements when giving a general characteristic of the person is mechanically justified by the fact that this ratio is proportional to l^2, i.e., the surface, and any energy penetrates through the surface. Maybe, the same ratio can be applicable for trees, seeds, fish, etc.

Inheritance, adaptability, and changeability of a form are considered to be fundamental properties of evolution. The mechanical principle of homogeneity and mechanical similarity seem to play some role in the theory of heredity. Due to this principle, all the forces are subordinate to a single equation, implying that a process develops in a similar way. For instance, a child is similar to an adult, although not geometrically; this similarity is different in different parts, depending on acting causes. Moreover, two main Newtonian rules can be added to these speculations:

(1) We shouldn't search for other causes in nature in excess of those which are true and sufficient for explanations.
(2) For this reason, natural appearances of the same kind should be explained due to the same causes, if possible.

We recall the same rules when we say about mimesis, the imitative colour. A technician is far from being able to interpret these phenomena in terms of biology, trying to explain the sameness and similarity in coloring simply due to the same causes. If butterflies and flowers have similar coloring, this seems to originate from the fact that the same reasons act in both cases. In any case, a common explanation seems to be irresponsible, not providing any consequences, as well as insufficient because it is unproved and impossible for us to believe that insect's eyes are able to perceive the beauty and tints in coloring. Interpretation in terms of mechanics, however, brings about a question what is the reason of the same coloring.

The idea of struggle for existence is distributed over all phenomena. It seems from the mechanical point of view that we should confine ourselves to the law mentioned above, namely, only those events occur which can happen under given conditions, with the least deviation from a free development. If a tree grows in overcrowded conditions, it does not struggle against other trees; instead, it develops in a way that is only possible under given conditions, with the least deviation from a free growth. The mechanical interpretation involves the comparison between the trees growing in a forest and in freedom and the determination of the least deviation in this case.

3 Changeability of a Form of Mechanisms

Finally, the idea of the changeability of a form can be considered from the mechanical point of view. The changeability of a form should be attributed to the dynamic action of forces rather than random causes. When a body is in rest or it moves uniformly, the action and counteraction are equal to each other, i.e., they are of static nature. If the motion is nonuniform, there is no equality between a driving force and a force of resistance. Since the processes in animated nature are controlled by dynamic forces because there is no constant accordance between surface and volume, the results should vary but oscillate around a mean value, a standard, which is thus protected by nature.

Goryachkin believed that in mechanics, the theory of changeability of forms can be most easily followed using mechanisms, which are very changeable and have complicated and intricate shapes. "Mechanisms are flowers of technics. The theory of mechanisms is a part of technics, which is the best elaborated among all others. Everything is clear and certain here" [1].

First of all, it should be noted that only three kinds of motion are possible in our space of three dimensions: translational motion along a straight line, rotational motion around an axis, and screw motion. According to this, all mechanisms can only consist of three pairs: (a) prismatic pair, which consists of a slide and a guiding straightedge; (b) revolute pair in the form of an axle with a wheel; (c) screw pair in the form of a screw with a nut.

Further Goryachkin wrote: "The construction of mechanisms only involves combinations and developments of these pairs. The whole task of an inventor is to make up new combinations. For instance, if only prismatic pairs are combined among themselves, this results in a mechanism of three prismatic pairs in a plane and a mechanism of four prismatic pairs in a space of three dimensions. If only revolute pairs are combined, a four-bar mechanism consisting of four revolute pairs in a plane and a mechanism of seven revolute pairs in a space of three dimensions may be obtained. Using screw pairs, a mechanism of three screw pairs in a plane and a mechanism of seven screw pairs in a three-dimensional space can be made up" [1].

A planar four-bar mechanism is the most popular (Fig. 3); it can be varied in many different ways.

Fig. 3 A planar four-bar mechanism

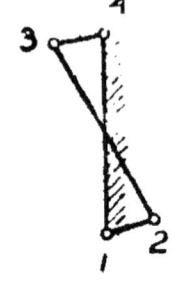

Thus, a lot of different *variations* of this mechanism can be obtained, when changing relative dimensions of links. If one revolute pair is substituted by a prismatic pair, a well-known mechanism of a steam engine, a crank with a conrod, is obtained. This mechanism is a *hybrid* of a four-bar mechanism.

Finally, the same mechanism can lead to *mutations*; a transition in this case resembles a jump, which occurs on the base of specific ideas. For instance, we may use an idea of transition from discrete motion to continuous motion. For this, two end levers should be replaced by wheels, or pulleys, while the intermediate one should be replaced by a belt; this construction is known as a pulley drive (Fig. 4).

Further, if a pencil is fixed to the belt, it will draw curves in the plane of both pulleys (evolvents of circles). These will touch each other all the time, resulting in a gear mechanism (Fig. 5). Therefore, a pulley drive and a gear drive can be considered as mutations of a four-bar mechanism, which are obtained as a result of continuous development of a four-bar mechanism, although in its own specific way each time.

In turn, the acquaintance with structures of mechanisms in animated nature, i.e., structures of plants and mechanisms of live movers, should be very fruitful for mechanics.

Fig. 4 Pulley drive

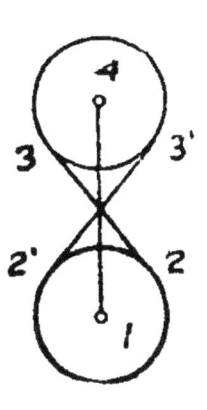

Fig. 5 Gear mechanism

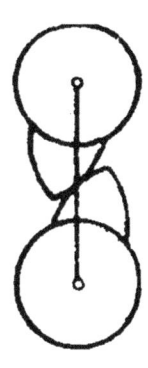

4 Live Movers and Mechanical Tractors

In many cases, such appliances and modes of operation exist in animated natures which are yet unreachable in modern technology. For instance, a human being, and live movers in general, may enhance their efficiencies by a factor of 10 or 20, whereas the energy store of technical items is confined within considerably lower limits. The ability of human beings to enhance their efficiency helps them in many cases, for example, when climbing up from deep chasms, whereas even a shallow pit makes a powerful tractor helpless. The analysis of walking of a human being allows us to conclude that the leg mechanism can be reduced to a four-bar mechanism (Fig. 6). It becomes completely clear if we imagine how a child moves on one skate, when one leg is put forward while another serves for pushing, i.e., it is a lead.

However, this method is only applicable until the moment when a person encounters an obstacle; when this happens, the person changes the mechanism, bending the knee of the leg that was earlier in front. Thus, the leg in front becomes a lead while the leg at rear serves as a support (Fig. 7). These processes repeat continuously in ordinary walking. The blind imitation of this method is not reasonable for technical constructions because animated nature cannot use wheels.

Levers should be replaced by wheels to effectively reproduce a mechanical model of legs of the human being.

To do this, we should put wheels instead of the levers 1–2 and 3–4, while the lever 2–3 is regarded as a frame (Figs. 6 and 7). In order to reproduce the device of passing through obstacles that is used by a human being, the following tractor can be built (Fig. 8).

Fig. 6 The scheme of the human legs as four-link mechanism

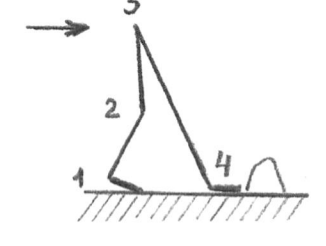

Fig. 7 The mechanism of the legs of a human in front of the obstacle

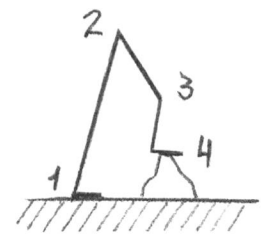

Fig. 8 The scheme of caterpillar tractor

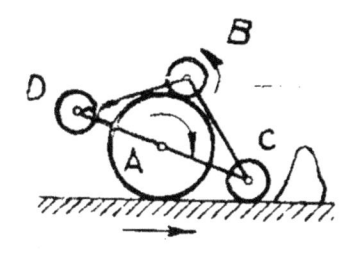

The leading wheel *A* (which replaces the leg 1–2 in rear) is driven into rotation using the cog-wheel *B* fixed to the frame.

At the frontal part of the frame, there is the wheel *C* that replaces the leg 3–4, which is put ahead in walking. Therefore, motion is accomplished using the leading wheel *A*, while the frontal wheel *C* serves as a support. At the moment when the wheel *C* comes to an obstacle, the tractor (similar to what a human being does) discontinues its motion, the cog-wheel *B* begins rolling over the motionless wheel *A*, in consequence of which the wheel *C* rises up and the wheel *D* goes down (Fig. 9). As soon as the wheel *D* touches the support surface, the tractor will begin moving again; the wheel *A* approaches the high spot and overcomes it easily, whereupon the tractor falls down and comes to the initial position anew. Goryackin mentioned that this system the 6-wheel tractor, as described in *Agricultural Mechanics* in 1919, was applied in England in 1927. He stated further that this model could be made intellectual to some extent, so that it could escape pits. For this, a wheel with a load should be put at the front; when approaching the pit, the load goes down, hindering the left wheel from moving forward via a transfer mechanism and forcing the tractor to turn aside.

Modern crawler tractors instinctively reproduce the next model. Imagine a cart *A* on which a person lies down, the body flat on the cart and the arms holding the wheel; by pushing away with a leg, this person moves ahead (Fig. 10). These remakes of live movers into mechanical engines may be considered instructive enough.

In order to reproduce a crawler tractor, it is sufficient to replace the lever 1–2 by a wheel, putting another supporting wheel in the front (Fig. 11). When encountering an elevation, the tractor lifts its front part in the same manner as a person on the cart does, passing through the obstacle.

Fig. 9 The scheme of a caterpillar tractor in front of obstacle

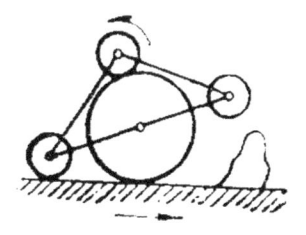

Fig. 10 Person is on the cart and holds the wheel

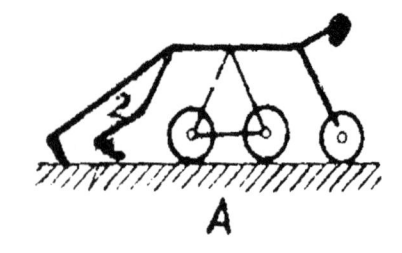

Fig. 11 The scheme of crawler tractor with wheel in the front

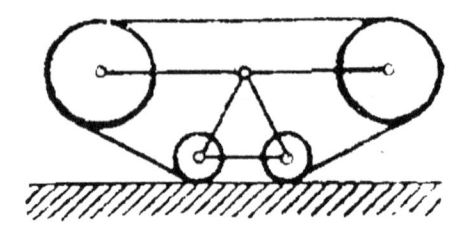

Depending on working conditions, it may be needed for a person to walk barefoot or in footwear, sometimes with spikes; to move on skis or on skates; to put planks on squashy ground, etc. Similarly, the rim of a tractor wheel cannot be of one and the same design suitable for all conditions. For different tractors, rims should be designed with regard to different conditions [2].

For firm "Hydropath" works, a mover to imitate the movement of a hamster in a wheel is used (Fig. 12). The mover F is hanged on the axle of the big wheel B, forcing to move the cog-wheel C connected with the wheel B. When rotating, the cog-wheel C begins to rise up along the wheel B, raising up the whole mover. As soon as the mover's centroid shifts beyond the fulcrum of the wheel, the mover's weight forces the wheel to rotate and move ahead.

Fig. 12 The engine to reproduce the movement of a hamster in a wheel

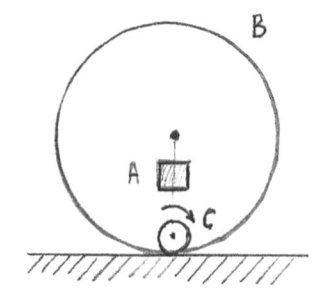

5 Conclusions

All this allows us to conclude that all the sciences should be guided by a united logic. Although we are far from claiming that all biological questions may be reduced to mechanics, the importance of mechanics for biology is beyond any doubt. According to Goryachkin, the basic method of mechanics involves the fact that mechanics starts from choosing a sufficient and necessary number of conditions for solving problems, schematizing a phenomenon and symbolizing it in the form of mathematical signs. After an equation has been composed, this leads necessarily to some consequences that may be used for checking the conclusions to either confirm or reject the solution. Otherwise, basic statements expressed in words often make conclusions uncertain and inconsistent.

Paper [1] ends with a conclusion, which is expressed by Goryachkin rather poetically: "Currently, natural sciences show a tendency for mathematics. The use of theory of correlations is spread more often than necessary. This method spreads to such an extent that only a correlation equation is often given, whereas experimental figures are omitted. It is impossible not to regret about this because the direct construction of digital data in the form of graphs provides a better view, having not only conditional value. Disadvantages of the correlation method become worse in "extrapolating", when the spread of conclusions is beyond the scope of measurements, which is especially unacceptable from the point of view of mathematics. In general, it is not reasonable to use only templates of mathematics when solving biological problems; it is necessary to involve also mechanics, the science on motion, that is, life."

References

1. Goryachkin V.P.: Mechanics of Animated Nature. Speech at the annual meeting devoted to the foundation of the Timiryazev Agricultural Academy (former Petrovska Academy), November 21/December 4, Moscow, pp. 1–8 (1927) (in Russian)
2. Goryachkin V.P.: Mechanical Tractors and Live Movers. In: Goryachkin, V.P. (ed.) Collected Works in seven volumes, Vol. 2. Moscow, Selkhozgiz, pp. 224–225 (1937) (in Russian)
3. Goryachkin V.P.: Operation of live engines, Moscow, Agronom (1914) (in Russian)

The Transformation of the Largest Aircraft Factory of Romania in Tractors Factory as Result of the Soviet Occupation

H. Salcă and D. Săvescu

Abstract After recalling the creation in 1925 and the rise of "Romanian Aeronautical Industry" (IAR) the most efficient plant in Romania during the period between the world wars this paper analyzes, in detail, its two periods of profound upheaval. The first in the context of the occupation of Romania by the Soviets and the establishment of the communist dictatorship sees the transformation of the IAR in tractor plants. The second, after December'89 in the process of exit from communism, is characterized by a long agony of the factory that ends with its liquidation. Until the end of the World War II, IAR produced more than 1,200 aircraft with over clean design and the other half under license: PZL (Poland), Fleet (US), and Savoia-Marchetti, Nardi (Italy), Fiesler-Storch and Messerschmitt (Germany). The Convention of the armistice with the Soviets was very strict and was only the beginning of a series of crimes (liquidation of elites) and abuse. Of these confiscated as compensation (followed by the nationalization) of certain industrial, removal or change of use to others etc. The latter case is illustrated perfectly by the IAR. In 1946, it changed its specialization, into producing tractors. The first was the IAR-22, a hybrid between Hanomag and Lanz Bulldog, followed by tracked Soviet models, KD and KDS. Early in 1960, appearing on the market the first fully Romanian tractor design, as well equipped with Fiat engines. The plant has grown, reaching a production capacity of 32,000 tractors per year, with 24.000 employees, having a good price on the market (11.000 USD in comparison with others, e.g. John Deere—25.000 USD). After December 1989, the situation of the plant was constantly deteriorated and often, employees of the factory took to the streets to express their discontent. In 2002, the company was still producing 4000 tractors. In 2004, it was close to the privatization; the potential buyer was the Italian

---

H. Salcă (✉)
Spiru Haret University of Brașov, Brasov, Romania
e-mail: horia.salca@gmail.com

D. Săvescu
Transilvania University of Brașov, Brasov, Romania
e-mail: dsavescu@unitbv.ro

Landini Group, or Mahindra. For different reasons (it's difficult to find a real reason) the privatization did not take place. In 2007 the plant was closed, entering in a process of liquidation and its assets were purchased by Flavus Invest Ltd. of Bucharest, owned by the British Investment Fund Capital Partener Centera.

Keywords Aviation · Romanian Aeronautical Industry—IAR · Brașov · Sovietization

1 Introduction

The Romanian Aerospace Industry of Brașov was the most efficient plant in Romania during the period between the wars. Founded in 1925, with the participation of French factories Blériot—SPAD, for cells and Lorraine-Dietrich, for motors, it began its activity in 1927 by the construction licensing aircraft Morane—Saulnier MS-35, 30 aircraft followed the next year by aircraft type Potez 25, a total of 280 aircraft.

During 20 years of existence, the Brașov plant would produce more than 1,200 aircraft, of which roughly half of own design, half of foreign licensing: PZL (Poland), Fleet (USA), and Savoia-Marchetti, Nardi (Italy), Fiesler-Storch and Messerschmitt (Germany). The maximum performance would be represented by the fighter IAR-80 considered at that time as the fourth world performance (speed of 510 km/h) after Messerschmitt Me 109 G (520 km/h), Hawker-Hurricane (570 km/h) and Curtiss-Wright P-37 (550 km/h) and in the area of the engine prototype IAR-7M, with 7-cylinder in star disposal, 370 HP to a tower of 2700 rotation/min, and a total displacement of 9.5 l [1] (Fig. 1).

The output of Romania from the alliance with Germany, and the remoteness of the power of Marshal Ion Antonescu, through the coup of August 23, 1944, meant —sadly—not only the country's occupation by the Red Army but also the beginning of the Sovietization of Romania (the Soviet will-power imposed in Romania).

The Armistice Agreement imposed strict conditions, which meant only the beginning of a series of abuse hard to imagine, which consisted in the plunder by industrial possessions confiscated as compensation (followed then by nationalization), the removal of some companies and the change of specialization for others— the case of the Romanian Aeronautical Industry Brașov—the persecution of intellectuals and prestigious specialists by crimes, convictions, deportations, confiscations and intimidation of any kind.

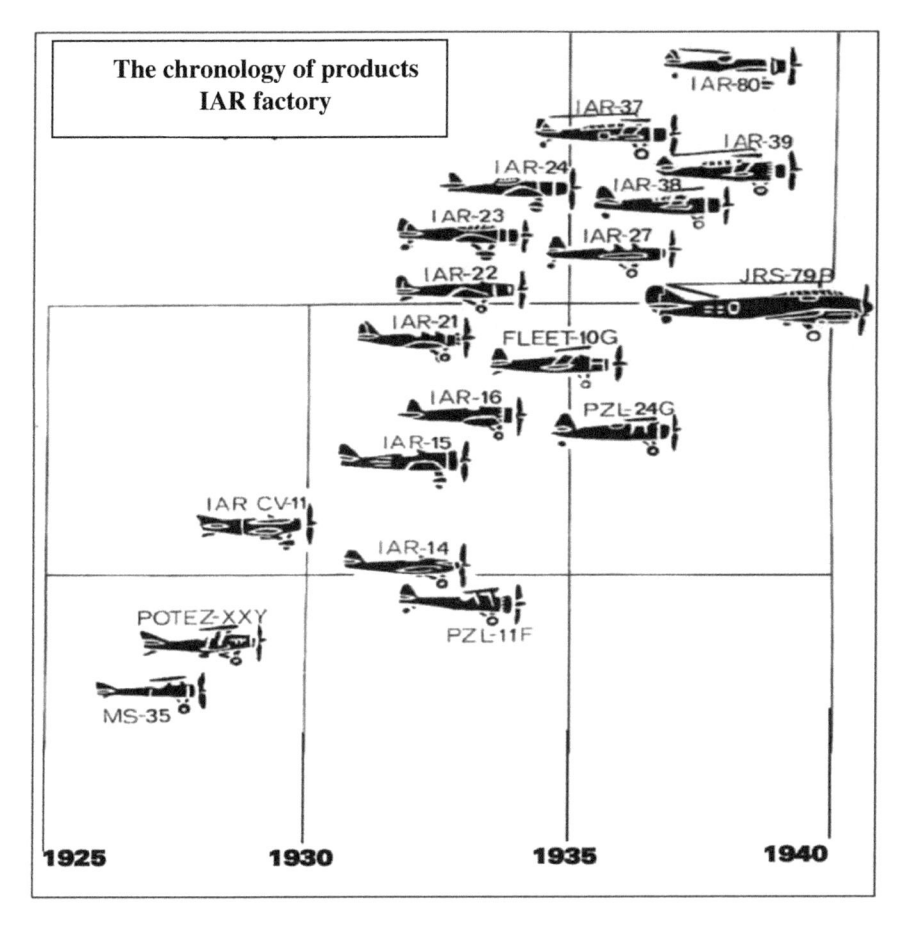

Fig. 1 The chronology of aircraft produced in Brașov

2 Background and Analysis

The Allies heavily bombed the territory of Romania and especially Bucharest and the oil zone around Ploiești, about which the British Prime Minister Sir. Winston Churchill said "*it represents*—no more no less—*the taproot of German power*". The April 12, 1942 took place Halpro mission, the first American raid over Ploiești which proved a total failure. A year later, August 1, 1943 the Tidal Wave operation, the longest raid since the beginning of the war of 7000 km, was going to be a very serious blow to the cities of Ploiești and Câmpina. On April 4, 1944, Bucharest was struck by the Anglo-American Aviation, which left behind over 3000 dead and 2000 injured, for the most part civilians, and 1 day later, Operation Overlord was again the objective of the city of Ploiești. The landing in Sicily and Calabria monitoring progress towards the north of the American troops and the occupation of

Foggia airports have created the premise of crossing the Carpathians to Allied aircraft. The day of Easter of 1944, April 16, at 10 h 30, American Liberator bombing squadrons, escorted by type of fighter P-38 Lightning bombarded the city of Brașov. They appeared over the city from the direction of the Massif Postăvarul, they crossed the city towards IAR and the train station, and then they withdrew towards Piatra Mare Mountain and then returned to Postăvarul Mountain. The first bombing destroyed much of the aircraft factory and those who followed: one on May 5 and again successive straight to short intervals have proved catastrophic. Following the destruction of the IAR factory, following the transfer of what has been recovered, to 12 scattering centres, for the continuous production activity.

If Allies' bombing during Romania's alliance with the Axis [2, 3] have imposed rapid dispersion of the plant (its "hiding" in territory being so necessary for continued front production) not in the same way could regrouping of the departments be achieved. It would be interesting to establish a scale of guilt and complicity of intellectuals and scientists from IAR or Bucharest with the Communists (the Soviets!) which brought about the failure of this great local factory, beyond the interest of the Russians to cease production of aircraft and their engines, at least for a long period of time [4].

If we were to identify the main obstacles to reunification, others than those related to the arbitrary moment, in the first place would, without any doubt, the lack of funds and materials for cleaning rubble, reconstruction and repair of the hangars and the bombed departments. Secondly, mention should be made of lack of means of transport which would go either to front or the USSR, after having been confiscated by the Soviet Allied High Command, on account of the truce, and the alleged war debts [5]. The same thing happened to a large number of machine tools, industrial machinery, equipment, reference material and technical projects. Thirdly and finally, the ruin the national economy caused by the entry of the Red Army meant a major lack of raw materials and semi-finished products, necessary to resume normal activity. In this section, we would like to underline the two main references relevant for this state of being "Goods delivered to the Soviet Army" and "Goods confiscated by the Soviet Army" [6] (Fig. 2).

Fig. 2 IAR factories after the bombing of the Allies (8.5 Pct)

Law no. 268
Mihai/Michael I,
By the grace of God and the national will, the King of Romania, to all those present and will, health.

The Assembly of Deputies voted and adopted and We sanction, the following:

Law for the creation of joint stock companies: the Anonymous Society by Shares of Metallurgical Enterprises of the State; the Anonymous Society of Chemical Enterprises, Pyrotechnics and Confections of the State.

Article 1—The Romanian State, through the Ministry of Industry and Trade, established from the publication in the Official Gazette of this Act, two limited liability companies known as:

A. The Anonymous Society by Shares of Metallurgical Enterprises of the State;
B. The Anonymous Society of Chemical Enterprises, Pyrotechnics and Confections of the State.

Article 2—The capital, purpose, duration, location and rules for the management and operation of these companies are those contained in the articles next to the law, they are an integral part.

Article 3—The State brings into:

A. The creation of the Public Limited Company Limited by Shares Metallurgical Enterprises of the State, all property of any kind, which includes the assets of the following company:

1. Iron Mills of the State of Hunedoara.
2. The Romanian Aeronautical Industry.
3. Margineanca plants.
4. The Romanian Arsenal.
5. The Arsenal of Sibiu.
6. The Plant for Protection Materials.
7. The Commercial Administration of Military Facilities of the Air Force.
8. The Commercial Administration of Marine Industrial Establishments.

This law and the statutes annexed were voted by 188 votes against two majorities.

Vice President, P. Constantinescu-Iasi

Secretary, Dr. Ştefan Cleja

We promulgate the law and order that it be given the seal of the State and published in the Official Journal.

Given in Bucharest July 12, 1947

Signatures,

Mihai/Michael I.

Minister of Industry and Trade, Gh. Gheorghiu-Dej

Minister of Justice, Lucretiu Patrascanu

3 Tractors Production

As a result of this law, IAR Brasov, in 1946 changed its specialization, producing tractors, having the name *Intreprinderea Metalurgica (Metallurgical Enterprise)*. In 1948 the plant became *Tractorul Factory of Brașov*. The both trade mark of airplane Factory and tractors Factory are presented in Fig. 3.

It is the moment of creation of Mixed Plants, a partnership between the Soviet occupation and the Romanian State—SovRom. They where mixed commercial society established in Romania at the end of the World War II and existed until 1954–1956. Between the two countries was signed on May 8, 1945, in Moscow, an agreement on establishing this organizations. Theoretically they should generate revenue for reconstruction after the devastation of the war. In reality they functioned as means of providing resources for the Soviets, further weakening the Romanian resources after being forced by the Soviets to pay severe war damages imposed by Peace Treaties of Paris, in 1947, the damage claimed by the USSR totaling 300 million dollars. Soviet contribution to SovRom companies consisted mainly in the sale of military equipment left the German troops on the battlefield, overvalued equipment every time and paid by Romania. Goods sent from Romania to the Soviet Union were estimated at $ 2 billion USD, far exceeding the damage demanded by Soviets, and until 1952, 85 % of Romanian exports were redirected to the Soviet Union. Last SovRom was dissolved in 1956 (Fig. 4).

The first tractor was the IAR-22, followed by tracked Soviet models, KD and KDS.

IAR 22 (Fig. 5) is the first Romanian tractor, model designed and produced in the plant Tractorul Brașov at the beginning of 1946, leader of project was professor Radu Emil Mărdărescu. It was a hybrid between Hannomag model and [4, 7, 8].

Lanz Bulldog, and the proposal was from engineer Ion Grosu. Machine weights was 3.4 tons, it was equipped with a Diesel engine of 38 HP. The force developed by engine was 1,225 kgf. The tractor was equipped with a manual gearbox having

Fig. 3 The two symbols: IAR (1925) change in UTB (1948)

Fig. 4 Transforming aircraft hangars in the production units of tractors

Fig. 5 The first tractor produced in Braşov: IAR-22

Fig. 6 IAR-23

3 + 1 speed, and a dry clutch. From the beginning, the plant received an order from the state consisting in 5,000 pieces. The first assembled tractor left the factory on 26 December 1946 and the tractor no. 1000 was completed in early 1949.

The next model, IAR 23 is an agricultural tractor and it has the same engine like IAR 22. PTO device, equipped with transmission shaft and belt pulleys, can set in motion other cars like the threshing machine, fan, combine, hardwood, etc. (Fig. 6).

In parallel where built crawlers under Soviet licence KD-35 and KD-35.2 (see Figs. 7, 8 and 9) [9].

Fig. 7 The caterpillar KD-35 in front of the headquarters of Sovromtractor Brasov

Fig. 8 The LTZ KD-35 tractor, 1947

Fig. 9 The KD-35.2 tractor, 1948

UTB KD-35 (tracked tractor) [9] equipped with a diesel engine UTB D-35-M (4 cylinders, 37HP) gearbox 5 + 1 gears. It had a coupling device that may tow agricultural machinery and was provided with secondary transmission shaft for driving active organs of agricultural machinery towed as: harvesters, mowing machines, binding machines, etc.

Early in 1960, appearing on the market the first fully Romanian tractor design, as well as models equipped with Fiat engines (Fig. 10). The plant has grown, reaching a production capacity of 32,000 tractors per year, with 24000 employees.

In all his history the UTB Plant from Brașov built a lot of types of tractors, as seen in Fig. 11.

Fig. 10 First Romanian tractor, U650, having Fiat engine

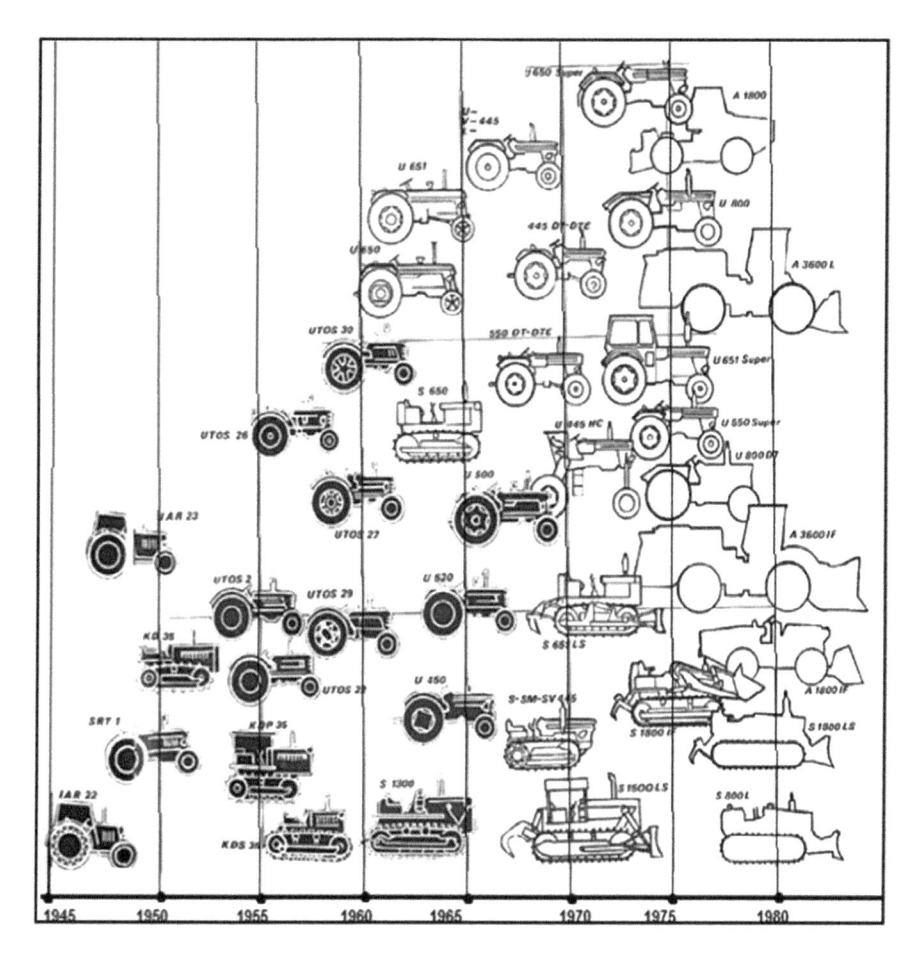

Fig. 11 The chronology of tractors produced in Brașov

4 After 1989

After December 1989, the situation of the plant deteriorated constantly and often, employees of the factory took go on streets/strikes to express their discontent. In 2002, the company was still producing 4000 tractors. In 2004, it was very near to the privatization; the potential buyer is the Italian Landini Group. But there where also others offer, e.g. Mahindra—India in 2005. For different reasons, difficult to discuss, the privatization did not take place. In 2007 the plant was closed, entering a process of liquidation and its assets were purchased by Flavus Invest Ltd. of Bucharest, owned by the British investment fund Capital Partener Centera.

Table 1 Employs evolution

Year	1980	1990	2003	2004	2005	2007
Employs	30.000	22.000	6.800	5.000	3.300	1.900

Fig. 12 Evolution of employs 1980–2007

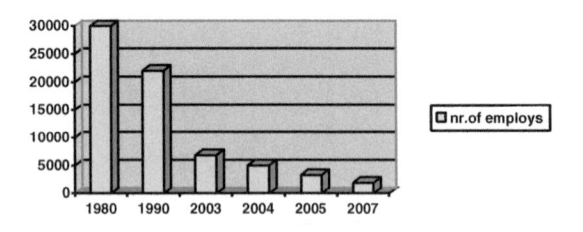

The "evolution" of number of employs between 1980 and until the end (2007) is presented in Table 1.

Is relevant what the downfall of the plant was in the last years. May be it was a problem of quality, reliability, obsolescence, but a tractor is a tractor.

Also, in Fig. 12 it is presented, as a diagram, the evolution of employs number.

5 Conclusions

After the August 23, 1944 events and the signing of the Armistice Agreement with the Allies, the Soviets took control over the whole of Romania, all the country economy was under control. Both, the Armistice Agreement and the Peace Treaty between Romania and the Allied and Associated Powers, signed in Paris February 10, 1947, limited the number of aircrafts for the military and banned aircraft production in Romania [10]. It was therefore the need converting the IAR Brasov factory moving the production from high performance airplanes to tractors.

It is good to known what it could happen if an airplane has damage in air or a tractor has a damage in the furrow: "airplane-flies and the tractor—plough" (in Romanian language: avionul zboară, tractorul ară).

Tractor production grew slowly, reaching the same level as impressive as volume and such as results. The 1989 Revolution, give us a chance to rebuild the economy oriented to market, but, in fact, was the ultimate failure of this industrial fortress of Brașov.

The last survivor IAR-22 tractor exists in the same place where it was built, the platform of IAR and UTB Plant but in an exhibition near a Mall (see Fig. 13).

Fig. 13 The last survivor: The IAR 22 tractor

References

1. Salca, H.: O istorie a Uzinei I.A.R. Brașov (A short history of Romanian Aeronautical Industry Plant Brașov), Ed. Universității Transilvania Brașov (2005)
2. Stănescu, F.: US bombers strike Bucharest in New York Times, vol. XCIII, nr. 31514, p. 1, Saturday, 6 May 1955
3. Stănescu, F.: Balkans blast lane in Italy to food fol, in New York Times, vol. XCIII, nr. 31514, p. 1, Saturday, 6 May 1944
4. Ardeleanu, I.: 23 August 1944. Documente. Convenția de armistițiu între guvernul roman și guvernele Națiunilor Unite (23 of August 1944. Documents. The Armistice Agreement between Romanian government and United Nations governments), vol. II, Ed. Științifică și Enciclopedică, București pp. 707–711 (1984)
5. Kirk, R.: Forțele sovietice au ocupat România (Soviet Forces occupied Romania). Macmillan Press (1990)
6. Stănescu, F.: Ocupația sovietică în România – Documente 1944–1946 (Soviet occupation in Romania—Documents 1944–1946), Ed. Vremea (1998)
7. Eisenhower, D.: Cruciada in Europa (Cruciade in Europe), Ed. Politica, Bucharest, pp. 232–233, 265–269 (1975)
8. Lache, St., Tutui, Gh.: România și Conferința de Pace de la Paris din 1946 (Romania and the Peace Conference, Paris 1946), Ed. Dacia, Cluj-Napoca pp. 286–348 (1978)

9. Roşulescu, V.: Tractorul românesc – scurtă evoluție (Romanian tractor—short evolution) (2012)
10. Răceanu, M.: România Versus the United States: Diplomacy of the Absurd, 1985–1989, p. 2. Palgrave MacMillan (1994)

New Trends in Learning Through 3D Modeling of Historical Mechanism's Model

O. Egorova, K. Samsonov and A. Sevryukova

Abstract The aim of this work is a detailed analysis of "Ferguson's planetary train", the model that represents a part of a scaled copy of the Solar system model. The peculiarity of the mentioned planetary model is that the rotational movement of input link of the mechanism is converted into the forward movement of its output link. A detailed analysis of so-called "Ferguson's Paradox" was conducted as well as the biography of James Ferguson was recounted. The created 3D model demonstrates the Ferguson's Paradox. Possible applications of the mentioned "paradox" in technical devices, engineering solutions and teaching process are considered.

Keywords Ferguson's paradox · Planetary mechanism · History of MMS · 3D modelling · 3D studio max

1 Introduction

At the end of the 18th century and later, the need for qualified, technically educated people appeared in Europe because of the industrial revolution and fast development of science and technology. New machines were in great demand as well as specialists able to operate them. The most important disciplines—Theory of Mechanisms and Machines (TMM) and its branches became the integral and necessary part of teaching future engineers in technical universities. The theoretical study of fundamental laws of sciences was combined with practical training for

---

O. Egorova (✉)
Theory of Mechanisms and Machines Department, BMSTU, Moscow, Russia
e-mail: tmm-olgaegorova@yandex.ru

K. Samsonov · A. Sevryukova
Rocket and Space Technology Department, BMSTU, Moscow, Russia
e-mail: sams1@bk.ru

A. Sevryukova
e-mail: alexandra.sevryukova@gmail.com

© Springer International Publishing Switzerland 2016
C. López-Cajún and M. Ceccarelli (eds.), *Explorations in the History of Machines and Mechanisms*, History of Mechanism and Machine Science 32, DOI 10.1007/978-3-319-31184-5_13

future engineers. That's why the scaled models of various mechanisms began to be used in teaching process.

In the 17th century, in Paris, the "Académie des Sciences" put forward the idea of creating a collection of models to be used in teaching engineering disciplines. The idea resulted in successful development of similar mechanism collections in other European universities. They included all possible types of tools, instruments and mechanisms that appeared due to the latest in those times inventions and discoveries [1].

The models were made in their full sizes or were reduced. They provided firsthand view of mechanisms and objects, everyone could see, feel and touch them. At the end of the 19th century, the well-known German mechanical engineer and a lecturer of the Berlin Royal Technical Academy Franz Reuleaux (1829–1905) created the first academic "Collection of kinematic models" with more than 800 exhibits. Later he became the President of the Academy. He was often called the father of kinematics. He directed the design and manufacture of over 300 beautiful models of simple mechanisms. Copies of models from this collection were used in Germany and in other countries as well [2].

2 Russian Mechanisms Collection at Bauman Moscow State Technical University

On June 18, 1863, the Emperor of Russia Alexander II (1818–1881) signed the General Charter of the Imperial Russian universities [1]. In accordance with the signed document, each University in Russia had to establish a "Collection of mechanisms" that was called sometimes a "Practical Mechanics Room", or a "Museum of Mechanisms" or a "Machinery Room". Additionally a new position of a "Keeper of Museum" was introduced.

That was the beginning of the Russian Mechanisms Collection at Bauman Moscow State Technical University.

The higher engineering and technical education in Russia began to develop a little later than in Western Europe, this development being under the strong influence of European experience. Young Russian scientists who should have become future professors of Applied Mechanics were trained at the best universities in Switzerland, Germany and France. Alexander S. Yershov (1818–1867), Fedor E. Orlov (1843–1892), Nickolay I. Mertsalov (1866–1948) and others were trained by Ferdinand Redtenbacher (1809–1863), Gustav Zeuner (1828–1907), Franz Reuleaux (1829–1905) and other well-known European engineers. All these scientists believed the collection of mechanisms' models to be one of the most important parts of the educational process.

Russian Mechanisms Collection at the "Theory of Mechanisms and Machines" (TMM) Department in Bauman Moscow State Technical University (BMSTU), established in the early 1930s of the 20th century, is one of the largest in the world.

The Collection includes more than 500 mechanism models and of them about 120 models that are related to the historical section.

The first exhibits were the models from the collections of Redtenbacher, Voigt and Reuleaux. They were brought from Europe. Later the Russian Collection expanded through the addition of models designed and manufactured by Russian specialists, and at present time, they form an essential and considerable part of the collection. It is important that the acquired models exist only in quantities of one to three copies in the world.

Of these, about 35 models are from the collection of Ferdinand Redtenbacher and were manufactured in the workshops of the Polytechnic school in Karlsruhe (Germany, mid-19th century). Many models were manufactured in Berlin in the workshops of G. Voigt in the 80s of the 19th century.

Unfortunately, it is not always possible to determine exactly who is the author or a maker of some models. The nameplates and labels can be found only on some models, so the attribution of models without logos was conducted in accordance with design features, external features and materials used. Some models from the Schroeder catalog are very close in structure to similar models of Redtenbacher. In addition, over the years, the models had repeatedly been subjected to repairs; some of the parts in them were replaced, which also makes it difficult to determine the authorship of a particular person.

It is necessary to take into consideration that makers themselves had made changes in the models that later were not described in catalogues. Therefore, the errors in the attribution of some models are inevitable. In the most complex cases, models are ascribed to one group or another conditionally.

Other models can be conventionally divided into three groups: educational models, models referred to the scientific school of Leonid N. Reshetov (1906–1998) and models referred to the scientific school of V.A. Gavrilenko (1899–1977). All scientific models are unique mechanisms made in accordance with inventions or research results. Reshetov's scientific school models are associated with the study of the structure of rational mechanisms, and Gavrilenko's scientific school models —with gear mechanisms. The third section consists of educational models or models of mechanisms that are studied in the course of Theory of Mechanisms and Machines: levers, gears, and cams. Some of these models were designed at the TMM Department of Bauman University and manufactured in its workshops.

The teaching of Applied Mechanics began in IMTS (Imperial Moscow Technical School—the predecessor of BMSTU) in the 40s of the 19th century. The first teacher was I. Balashev. He initiated the collection of models and founded the showroom for mechanisms. The IMTS mechanisms showroom was greatly enlarged by its director and teacher of Applied Mechanics Alexander Yershov. Under his leadership, students had created more than one hundred models of mechanisms. Some of them, as we noted earlier, were made on basis of Redtenbacher models, because their drawings had been published. It should be noted that A. Yershov not always tried to make an accurate copy of those models. While creating a model, many design elements can be simplified and cheap and available materials can be easily used (Fig. 1).

Fig. 1 a Alexander S. Yershov (1818–1867). **b** Nickolay I. Mertsalov (1866–1948) [3]

The criteria of similarity in models are structural, geometrical and kinematic similarities. Much less common are models with dynamic similarity.

3 Ferguson's Planetary Train

The model № 9, Ferguson's planetary train (Fig. 2), belongs to the Russian Mechanisms Collection at BMSTU. It consists of two identical wheels—gears Z_1, Z_3 that are installed on a planet carrier, a link of a planetary gear train on which the axles of the planetary gears are arranged. The model has also one smaller wheel—gear Z_2. By rotating the planet carrier, we can watch that a small plate attached to the gear Z_3 moves forward and the plate is constantly in a horizontal position.

Each gear is mounted on the planetary carrier with bearings. The mechanism is mounted on a tripod, but it also could be installed on a frame.

The detailed study and searching through technical literature confirmed the hypothesis that Model № 9, the Ferguson's planetary train, represents a part of a scaled copy of the Solar system model called the Orrery, which we consider in more detail below.

Fig. 2 Ferguson's planetary train from the Russian Mechanisms Collection

Fig. 3 James Ferguson (1710–1776) [4]

The original model was created in the 18th century by the Scottish astronomer James Ferguson (Fig. 3) to illustrate the direct motion of the Moon's apogee and the retrograde motion of its nodes.

3.1 James Ferguson

James Ferguson (1710–1776) was a self-educated son of a Scottish crofter (Fig. 3) [4]. The story is told of the young Ferguson's fascination with mechanics being inspired by watching his father use a large lever to adjust the roofline of their house.

In 1742 Ferguson built his first Orrery. The term "Orrery" is a kind of a misnomer—the word "Tellurian" is a better term for a teaching device focused on the Sun, Earth, and Moon. Nevertheless, the word Orrery is understood by almost anyone interested in popular astronomy, astronomical mechanical devices, scientific instruments, the history of ancient clocks and devices, etc.

By April 1748, James Ferguson became already a popular scientific teacher and lecturer. His later courses, delivered in the provinces as well as in London, covered a wide range of experimental science. Among his inventions (besides eight orreries) we can find a tide-dial, a 'whirling-table' for displaying the mode of action of central forces, the 'mechanical paradox,' and various kinds of astronomical clocks, stellar and lunar rotulas [5, 6].

Ferguson's paper "Astronomy explained on Sir Isaac Newton's Principles" was published in July 1756, and met with immediate and complete success [4]. The first issue was sold out in a year; the thirteenth edition, revised by Brewster, appeared in 1811, and the demand for successive reprints did not cease until 10 years later. It was translated into Swedish and German, and there were no other treatises on the same subject for a long time. Although containing no theoretical novelty, the manner and method of its presentation were entirely original. Astronomical phenomena were for the first time described in familiar language. Thus, besides being one of the first popularizers of science, Ferguson's influence extended considerably, with people studying his books.

James Ferguson designed orreries and a number of clocks. He knew Benjamin Franklin and apparently was inspired by Franklin's famous 3-wheeled clock, later designing his own versions. Ferguson was a very prolific inventor, making many devices for use in research and lecture demonstrations.

3.2 James Ferguson's Orrery

In 1764 James Ferguson built the Orrery (Fig. 4) [4] for use in his lectures. He wrote: *"This machine is so much of an ORRERY, as is sufficient to show the different lengths of days and nights, the vicissitudes of the seasons, the retrograde motion of the nodes of the Moon's orbit, the direct motion of the apogeal point of her orbit, and the months in which the Sun and Moon must be eclipsed."*

The Orrery is not just decorative—this is a valuable teaching instrument. You can follow Ferguson by showing people how the Earth maintains its orientation as it is orbiting the Sun, why seasons happen, and why summer and winter are reversed in different hemispheres.

The Orrery could demonstrate the movement of the Moon and to predict its positions, it illustrates how eclipses can happen when the nodes of the Moon's orbit intersect light coming from the Sun, and how the nodes and apogee of the Moon's orbit move over time. Ferguson's activity also supported the process of making scientific instruments and introducing measuring devices.

It is important that even today's modern technologies have not replaced people's natural curiosity about the Earth and Moon. The Orrery built in 18th century by James Ferguson looks fascinating and can be used nowadays to popularize and explain astronomy to pupils, students, science classes, and more.

Ferguson's Orrery illustrates the process of discovery and human general fascination about the heavens. It provides a good example of the interrelation between innovative thinking (invention) and precision engineering (execution).

Moreover, Orrery could demonstrate a motion of a gear train that was later called as Ferguson's Paradox [4, 7].

Fig. 4 Ferguson's Orrery. Illustration by L. Whittemore [4]

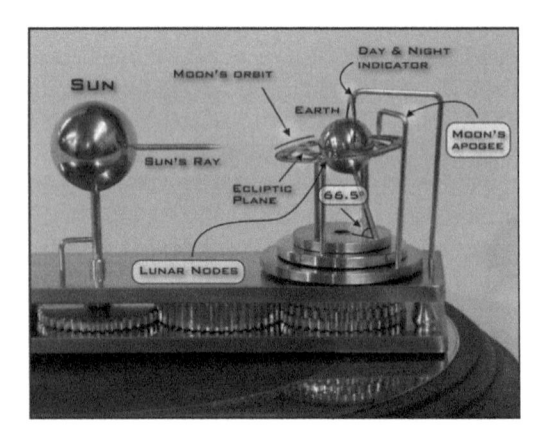

3.3 Ferguson's Paradox

Even today, "Ferguson's Paradox" amazes people. A simplified version of the paradox reads: "Three wheels on the same axis mesh with one thick wheel. Turn the thick wheel. One of the thin wheels goes forward, one backwards, and one goes no way at all!"

The "paradox" arises when in a train of gears—Z_1, Z_2, and Z_3—gear Z_1 is fixed and gears Z_2 and Z_3 have epicyclical motion around it [8–10]. Gear Z_1 is the gear "under the sun" (Fig. 4) and is fixed to the base. When all three gears have the same number of teeth, gear Z_2 rotates twice for each rotation and gear Z_3 maintains its orientation to a fixed frame of reference. That keeps the Earth's axis pointed in the same direction. When gear Z_3 has fewer teeth than gears Z_1 and Z_2, it turns in the direction opposite the mechanism, in this case illustrating the regression of the nodes. When the gear Z_3 has a few more teeth, it will slowly turn in the same direction as that of the mechanism, illustrating the advancement of the apogee of the Moon's orbit.

3.4 Structural and Kinematic Analysis of Ferguson's Mechanism

Kinematic analysis is the study of the movement of parts of the mechanism without considering the forces that cause this motion. When solving problems of kinematic analysis, all existing methods are used: graphic, graphic-analytical method (plans of velocities and accelerations) and analytical [11].

Let us conduct structural studies of Ferguson's planar mechanism. Consider a differential version of the mechanism where a planet carrier and all of the gears can rotate about their axes. Number of the moving links of the mechanism $n = 6$. Number of one-mobility revolute pairs $p_1 = 6$. Number of two-mobility higher pairs $p_2 = 4$. Mobility of the mechanism calculated by the Chebyshev's formula:

$$W^{pl} = 3*n - 2*p_1 - 1*p_2 = 3*6 - 2*6 - 1*4 = 18 - 16 = 2 \quad (1)$$

If we stop central gear $1(Z_1)$ in the mechanism then we will find a one-degree-of-freedom mechanism where $n = 5$, $p_1 = 5$ and $p_2 = 4$:

$$W^{pl} = 3*n - 2*p_1 - 1*p_2 = 3*5 - 2*5 - 1*4 = 15 - 14 = 1 \quad (2)$$

An equation which relates rotational movements of the links of the differential mechanism can be obtained from the Willis' formula [8]. We write this formula for all variations the gearing:

$$\frac{\omega_1 - \omega_h}{\omega_2 - \omega_h} = -\frac{Z_2}{Z_1}; \tag{3}$$

$$\frac{\omega_2 - \omega_h}{\omega_3 - \omega_h} = -\frac{Z_3}{Z_2}; \tag{4}$$

Then, for the gearing with gear 3(Z_3):

$$\frac{\omega_1 - \omega_h}{\omega_2 - \omega_h} * \frac{\omega_2 - \omega_h}{\omega_3 - \omega_h} = \frac{\omega_1 - \omega_h}{\omega_3 - \omega_h} = \left(-\frac{Z_2}{Z_1}\right) * \left(-\frac{Z_3}{Z_2}\right) = \frac{Z_3}{Z_1}; \tag{5}$$

$$\omega_1 * Z_1 + (Z_3 - Z_1) * \omega_h + \omega_3 * Z_3 = 0; \tag{6}$$

For Ferguson's mechanism where gear 1(Z_1) is stopped and $\omega_1 = 0$, angular velocity of gear 3(Z_3) is

$$\omega_3 = \left(\frac{Z_1 - Z_3}{Z_3}\right) * \omega_h \tag{7}$$

That is in our mechanism $Z_3 = Z_1$, then $\omega_3 = 0$ and gear 3(Z_3) makes circular translation [12].

We have built a plan of velocities of Ferguson's mechanism (case $Z_3 = Z_1$). Point "p" (Fig. 5) is the instantaneous center of velocity, the velocity which equals zero.

Given the above, it can be concluded that the "small plate" located on the wheel Z_3 of the studied Ferguson's planetary train (Fig. 2) will always be in a horizontal position.

Fig. 5 Scheme of Ferguson's planetary train and its kinematic study

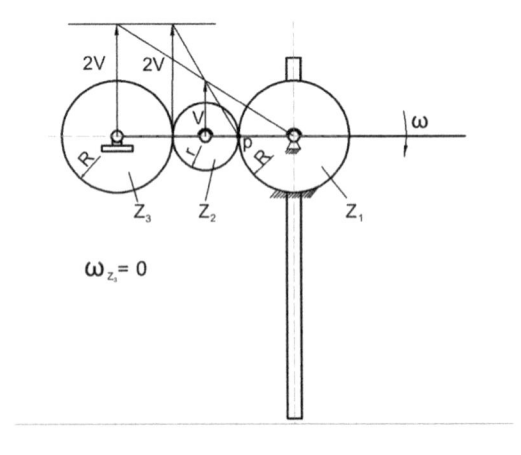

Fig. 6 Photos of Alexander Yershov's models from the BMSTU collection

3.5 Existing Analogues and the Use of Paradox

At TMM BMSTU Department in addition to the reporting mechanism № 9 there are other artifacts that can be called Ferguson's mechanism. These include labeled with number K-13(839) and K-14(93м) models of Alexander Yershov (Fig. 6a, b).

Models made by Alexander Yershov are similar to the Redtenbacher's models [13], but they have many differences in the shape of parts and their sizes. In the introduction to [14] it was noted that the models from the collection are divided into four classes: (1) Machine parts; (2) Typical engineering products; (3) Transferring and motion conversion mechanisms; (4) Models of machining complexes.

Models of Class 1 demonstrate correct structural forms and proportions of machine parts. Models of Class 2 are the results of some production processes. Models of Class 3 exactly show the real effect of different transmission mechanisms. Models of Class 4 represent complete machines and mechanisms, which are difficult to show by drawings.

Models K-13(839) and K-14(93м) belong to Class 3 and the structure and kinematics of these models are equivalent to "Ferguson's Paradox" model.

At present, Ferguson's Paradox is widely used in astronomy, engineering and even in the manufacture of watches. British blog on creation of wooden clocks "Brian Law's wooden clocks" put a plan on its website [15] through which you can make your own wood clock (Fig. 7), which clearly applies the "paradox". Using this Ferguson's Paradox, it is possible to design a gear train that will give a 12:1 reduction, so that we can drive the hour hand by the minute hand shaft. "This

Fig. 7 Brian Law's Wooden clocks "The Ferguson's Paradox Drive" [15]

simple model employs this technique to create a Teaching Clock that can be used to teach young children to tell the time from an analogue dial instead of reading from a digital display on their iPhone" [15].

4 3D-Modeling of Ferguson's Planetary Train

4.1 Advantages of Autodesk 3ds Max

The original model of the counting mechanism were photographed, measured and disassembled. Modeling started with individual parts of the mechanism.

Model elements in 3ds Max were made from primitive objects such as a box, cylinder, sphere, pyramid and so on. Individual elements of the part were created by using the rotation operation or moving primitive objects along their rail.

3D modeling of "Ferguson's planetary train" № 9 was carried out with the help of the program Autodesk 3ds Max (Fig. 8). This program has a high performance and provides a comprehensive modeling, animation, simulation, and rendering solution for motion graphics as well as for teaching purposes. Autodesk 3ds Max has grown to be one of the top 3D animation software options nowadays. It is developed and produced by Autodesk Media and Entertainment and has modeling capabilities, a flexible plug-in architecture and can be used on the Microsoft Windows platform. It delivers new efficient tools, accelerated performance, and streamlined workflows to help increasing overall productivity for working with complex, high-resolution assets. For students, teachers and educational institutions provide the ability to use Autodesk tools in full access is absolutely free. Of course,

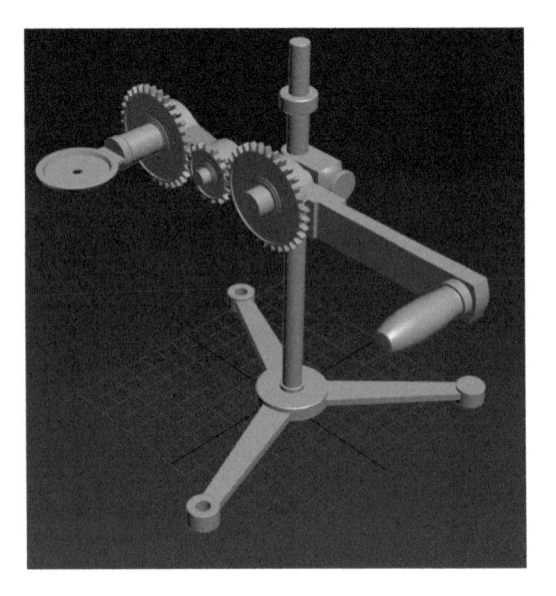

Fig. 8 3D model of Ferguson's planetary train from the Russian Mechanisms Collection

it is possible to use and another program for 3D modeling, but according to the authors of the 3ds Max the most optimal.

4.2 Modeling

Before we start modeling, let's divide the mechanism into parts. We will start with the rack (Fig. 9), for which we need to create a cylinder to control its size and click "Create". Then we need to add polygons in the "parameters", as well as to create an "editable poly" for it for further changes in our cylinder. After these operations we use "Extrude". So, we have a rack for our mechanism.

Then we make legs. Here we use "box", that is, an elongated cube, set its sizes, and click the "Create" button. We also create "editable poly" for it. Once created, we use polygons, that is, we identify the extreme points of a cube and narrow it, and this can be seen in the last picture below. Then we copy the legs and place them at an angle of 120° from each other, and connect them with our stand, using "Attach".

Then we make the mounting stand. We take three cylinders, in two of them we have to make holes for fasteners. These holes can be done using "Proboolean". To do this, we need to copy the cylinder and reduce the other internal cylinder and press "start Boolean", and get a hole.

In this picture, we make the holder (Fig. 9). We made it in three parts. First we made a "box", then to the right of it we add half of the cylinder, and the left side is slightly squeezed, using the known command "extrude", and smoothed with the command "smooth". Next we make cylinder and by using the "extrude", squeeze portion with a small diameter out of it.

Then we dealt with the driver (Fig. 10). We needed three cylinders and a cube (28 mm × 220 mm). We had to create "Editable poly" for all figures. We changed the cube using polygons and functions "Extrude" for what it would not differ from the actual mechanism. Next, we arranged cylinders and placed cylinder 2 at 15 mm below the others. At the end, we used "Attach" to connect the cube with the three cylinders.

After that, we proceeded to the gear wheels. We made them out of the cylinder, added 64 polygon edges, (because there were 32 teeth), similarly with gear 2, then

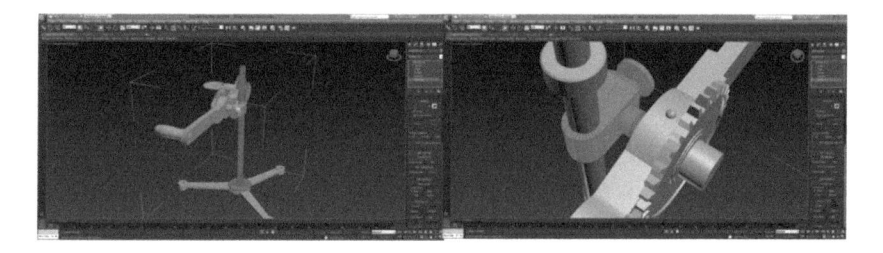

Fig. 9 Screenshots of the rack and the holder of 3D Ferguson's planetary train

Fig. 10 Screenshot the driver 3D model Ferguson's planetary train

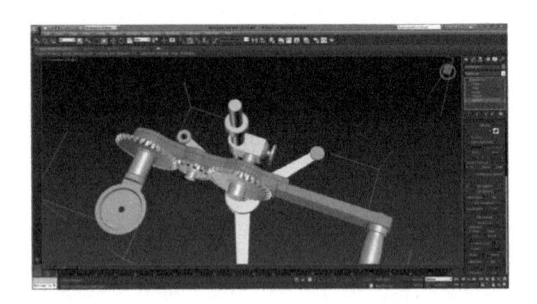

extruded them using "Bevel". Since we had bearings, they also had to be inserted; to do this, we needed to remove the extra polygons of the three cylinders to put our bearings and after a successful deletion, we inserted them and connected using "Attach".

Then we made a bearing. We had to make internal and external rings, they were made from the same cylinder, the function "Proboolean", that is, cut a small cylinder (smaller diameter) from the larger cylinder, then inserted a sphere.

We create a shaft by means of the cylinder and the function "Extrude". The handle is made of the same cylinder and by means of "Extrude". Then we expand the base, and process it with function "Smooth".

Next, we made a stand "bowl" of the "cone", and using "Proboolean" made a hole; on the right of the stand, we had a cube with altered polygons, and behind it there was the same cylinder. After their creation, we combined them with the functions "Attach".

Having completed all the above-mentioned steps, we obtained a 3D model, which is similar in appearance to the Ferguson's planetary train № 9 from the Russian Mechanisms Collection.

5 Animation

In developing the 3D model, we knew that it would be necessary not only to create a completely similar in outward appearance model, but it should also demonstrate the exceptional and unique mechanical properties, namely the Ferguson's Paradox. It was decided to create an animation (Fig. 11) and two videos, which would demonstrate the mechanism operation and the process of "assembling" the mechanism.

To create the animation of the mechanism operation, we needed to place the bones in 3 parts where real disposable shafts, included in the bearings gears, created them using "Link" and "Select".

Next, using the "Auto key" we set up the animation. The point was simple: as the number of teeth of the first and second wheel differ exactly as 2:1, set to rotation

Fig. 11 Screenshot animating 3D model Ferguson's planetary train

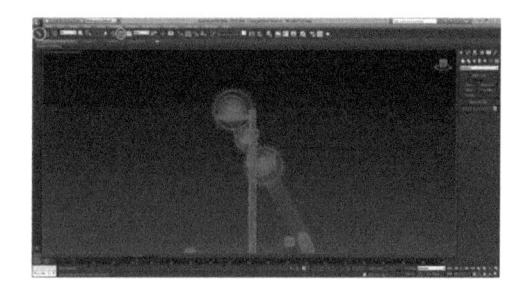

of the first wheel a full turn 360°, and the second wheel 720, 1 wheel motionless, and a plate we make so, that it would not revolve, and perform plane-parallel motion.

6 Conclusions

At the beginning of our work, we set a number of problems for ourselves, which have been solved in full:

1. In this paper, we considered the biography of James Ferguson in detail and disclosed the personality of this outstanding Scottish astronomer and instrument-maker of the XVIII century. Through this article, we tried to tell the reader the life story of James Ferguson, little known in Russia, who made an enormous contribution not only to astronomy, but also to the development of engineering.
2. Three-dimensional modeling has become an integral part of our lives. Due to 3D, we are able to create and animate realistic 3D model of the existing collection of BMSTU. The created 3D model demonstrates the Ferguson's Paradox. The model can be used for educational purposes in technical and humanitarian universities of different specialties. The created model will help in the preservation of the historical artifact of the BMSTU collection. The model eliminates the need for handling real models in classrooms and buildings (you need only a PC with the ability to output video or read the format (max)).
3. There are structural and kinematic studies of Ferguson's mechanisms. The kinematic analysis is decided to be used for creating a dynamic model, and later to implement the model not in 3D format but to make it from real materials. Simpler models without any problem can be printed on a 3D printer as described in detail in [14, 16], the method being now quite widespread.
4. The results of this work can be widely used in the teaching process [17]. Teaching technical disciplines, including TMM, using three-dimensional computer modeling, enhances professional skills and culture of a future specialist, motivating his need for constant self-improvement. New educational technologies are an effective method of teaching students and contribute to a better

understanding of studied subjects as well as to popularization and distribution of scientific knowledge.
5. In future, we are planning to create a dynamic model and to implement it using composite materials or composites with metal parts. At present, it is impossible to make the copy of Model № 9 from composite materials because of the lack of dynamic analysis [18].

References

1. Tarabarin, V.B., Carbone, G.: Collection of Bauman University: research school professors L. Reshetova and V. Gavrilenko. In: IFToMM World Congress, Besançon, June 18–21 (2007)
2. Redtenbacher, F.: Die Bewegungs-Mechanismen: Darstellung und Beschreibung eines Theiles der Maschinen-Modell-Sammlung der polytechnischen Schule in Carlsruhe/von F. Redtenbacher. F. Bassermann, Heidelberg (1866)
3. Moon, F.C.: Robert Willis and Franc Reuleuax: pioneers in theory of machines. Notes Records Royal Soc. Lond. **2003**, 209–230 (2003)
4. Carbone, G., Ceccarelli, M.: Experimental tests on feasible operation of a finger mechanism in the LARM hand. Int. J. Mech. Based Des. Struct. Mach. **36**, 1–13 (2008)
5. Henderson, E., Ferguson, J.: Life of James Ferguson, F.R.S. In a Brief Autobiographical Account, and Further Extended Memoir. Nabu Press (2013)
6. Ferguson, J., Editor Henderson, E.: Life of James Ferguson, F.R.S. In a Brief Autobiographical Account, and Further Extended Memoir. Cambridge University Press (2010)
7. Armstrong Metalcrafts. James Ferguson's Mechanical Paradox Orrery (2011–2015). http://armstrongmetalcrafts.com/Products/ParadoxOrrery.aspx
8. Golovin, A., Tarabarin, V.B.: Russian Models from the Mechanisms Collection of Bauman University. Springer (2008)
9. Downie, J.: Ferguson mechanical paradox motion work (2011). http://www.youtube.com/watch?v=vWNLDgU6Xl4
10. Mekanizmalar.com. Gears. Ferguson's Paradox (2015). http://www.mekanizmalar.com/fergusons_paradox.html
11. Tarabarin, V.B.: Theory of Mechanisms and Machines Lections. 15th Lection. Planetary Mechanism Kinematic (2010). http://tmm-umk.bmstu.ru/lectures/lect_15.htm
12. Orrery, B.T.: A Story of Mechanical Solar Systems, Clocks, and English Nobility (Series: Astronomers' Universe). Springer (2014)
13. Redtenbacher Collection of Kinematic Mechanisms, University of Karlsruhe, Germany. Model: UK053 Worm and Two-wheel Drive Chain (2012). http://kmoddl.library.cornell.edu/model_metadata.php?m=347
14. King, H.C.: Geared to the Stars. University of Toronto Press (1978)
15. Brian Law's wooden clocks. Projects Fergusons Paradox (2015). http://www.woodenclocks.co.uk/page72.html
16. Millburn, J.R.: Wheelwright of the Heavens. Vade-Mecum Press, London (1988)
17. Egorova, O.V.: Three-dimensional Computer Modeling in Teaching "Theory of Mechanisms and Machines" Discipline. Izvestiya vuzov. Mechanical № 7 (644), pp. 79–86 (2015)
18. Savchuk, I.: Blogerator's Litdybr Engine. 3D printing: the third industrial and digital revolution (2014). http://blogerator.ru/page/3d-pechat-industrialno-cifrovaja-revoljucija-3d-printer-makerbot-cena-opisanie-perspektivy-1

Some Inventions by Engineers of the Hellenistic Age

C. Rossi

Abstract Some examples indicating the surprising level of the technical and scientific knowledge of the Hellenistic scientists and engineers are presented. The latter concern the measuring of the time, the self-propelled carts, the throwing machines and the automatic devices. Some of them, in fact, already contain the concept of automation. A brief reference is also made to the steam cannon.

Keywords History of MMS · Hellenistic knowledge · Mechanics of classic age

1 Introduction

In modern historiography, the term Hellenistic period refers to the historical and cultural period starting with the military expedition of Alexander the Great in 334 B.C. in the East (or his death in 332 B.C.) and ends with the beginning of the Roman Empire (formally in 31 B.C.) when the last Hellenistic Kingdom (the Ptolemaic Egypt) was conquered.

To that period belong so many gigantic figures of scientists, engineers and inventors, whose knowledge was widely used in Europe till the 18th Century and can still surprise modern scientists and technologists (see e.g. [1, 2]).

It is reasonable to suppose that part of this knowledge was "imported" in Europe by the scientists that were in the Alexander's train. Nevertheless, in most cases, it seems more reasonable to consider an interaction of cultures between East and West than to mere transfer of knowledge.

C. Rossi (✉)
University of Naples "Federico II", Naples, Italy
e-mail: cesare.rossi@unina.it

2 Some Examples

2.1 The Water Clock by Ctesibius

Measuring of the time, during the classic age involved another problem. The length of a Greek or Roman hour was not constant since it was defined as 1/12 of the time between sunrise and sunset during the day and 1/12 of the time between sunset and sunrise during the night. Thus, time duration of 1 h was different from day and night (except at the equinoxes) and from a given day to another one. The water clock, designed by Ctesibius, solved this problem. A perspective reconstruction is shown in Fig. 1 on the basis of what was described by Rossi et al. [2], Vitruvius [3] and Russo et al. [4].

The main parts of the mechanism are shown, in an orthogonal section, in Fig. 2. The problem of measuring hours of variable length was solved by Ctesibius by fitting the dial on a shaft that was off the centre of the pointer shaft and by moving the dial during the year.

A bottom tank was filled by a constant water flow from a top tank that is constantly maintained full. A yarn was connected to the ball clock and to a counter weight and was wrapped in coil around the pointer axle. The bottom tank was drained daily and the cycle started again.

At any time the float passes through a certain position (once a day), it moves a rod that pushes one tooth of a gear. This last gear has 365 teeth, so it made a full revolution in 1 year, and was fitted on a hollow shaft coaxial to the pointer shaft and connected to a rod, as shown in Fig. 2. The dial was mounted on a hub having two orthogonal slots. Through the vertical slot passed the pointer shaft and in the horizontal one a crank was located and connected to the gear shaft. While the crank rotates, the dial could move just along the vertical direction. In this way, the dial

Fig. 1 Virtual reconstruction of the water clock by Ctesibius

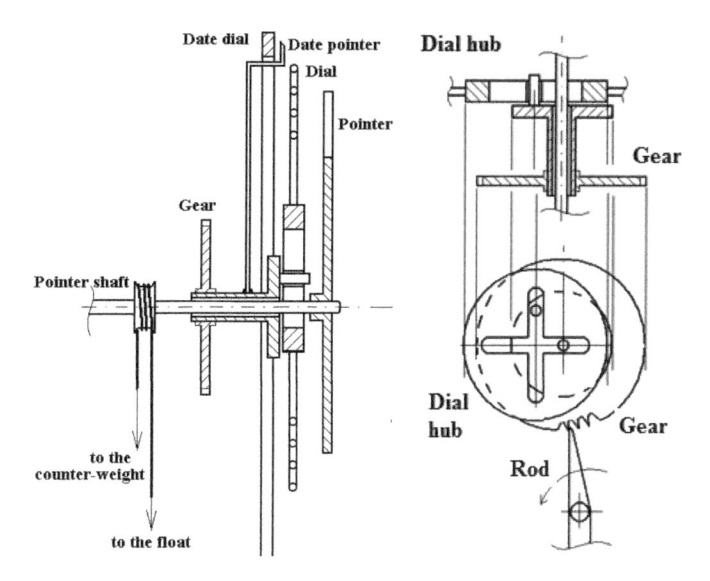

Fig. 2 Scheme of the mechanism for the dial motion

centre moves with respect to the pointer axis from the higher position to the lower position to the higher again, once in a year. The 365 teeth gear moved also another pointer to indicate the day of the year.

2.2 Heron's Self-propelled and Programmable Automata

Heron of Alexandria, in the 1st century A.D., designed and built automata representing animals in a sort of theatre in which the actors were automata moved by this kind of motor and by a device that permitted, among others, to program the law of motion of the automaton itself [5–11]. To this end interesting references can be found in the Heron's treatise Perì Automatopoetikes = about automatics [8, 9] in which, figurines mechanically moved (without any action from outside) in an automata's theatre, are described. The treatise by Heron was translated during the Renaissance by Berardino Baldi, abbot of Guastalla, (1553–1617) [10]. In this work he describes, among other things, some examples of mobile automata moved by a counterweight motor. In Fig. 3, on the left, is reported a drawing from Baldi's work: the counterweight, the rope linked to the latter and rolled on the wheels axle are clearly observable.

In the figure it is also possible to observe the third wheel that is idle and the counterweight that is located in a tank filled with millet or mustard seeds in order to regulate the counterweight motion. Also very interesting are the systems, invented by Heron and described by Baldi, to change the cart's direction.

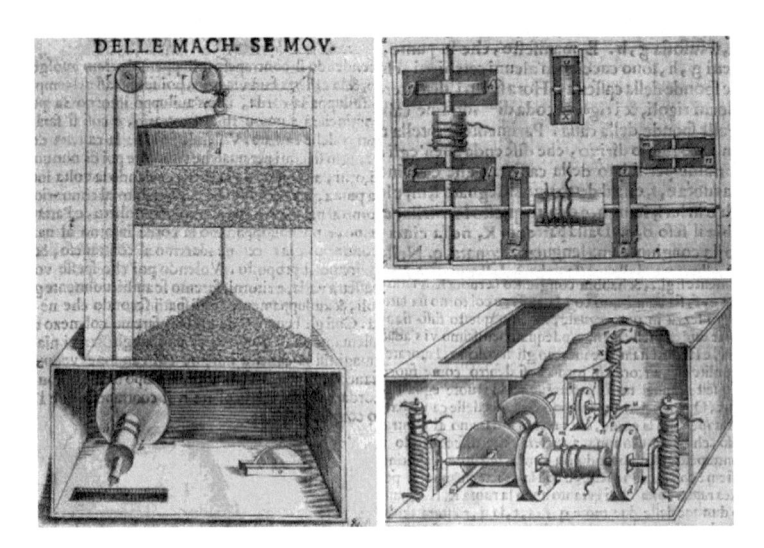

Fig. 3 Counterweight motor of Heron's self propelled automaton and mechanism to change direction [10]

In Fig. 3, on the right, there are two Baldi's drawings which show a first system used to change the cart's direction: in the drawing above it can be observed that two driving axles are used, each one is perpendicular to the other; in the same way, also the axles of the idle wheels are perpendicular. During operation, two driving wheels and an idle wheel stands on the ground while the other wheels (whose axles are orthogonal to the former ones) are lifted up. By means of screw jacks, shown in the lower drawing in Fig. 3 (right, below), that are also operated by ropes, it is possible to take down the wheels whose axle is orthogonal to the ones' that stand on the ground.

Another system to change direction seems even more interesting because it uses the programmability of motion concept; this system is also attributed to Heron and is described by Baldi [10]. In Fig. 4, on the left, another drawing from the work by Baldi is reported. The axle of the driving wheels is divided in two axle shafts that are independent one from the other; on each one of the latter a rope is rolled. If the

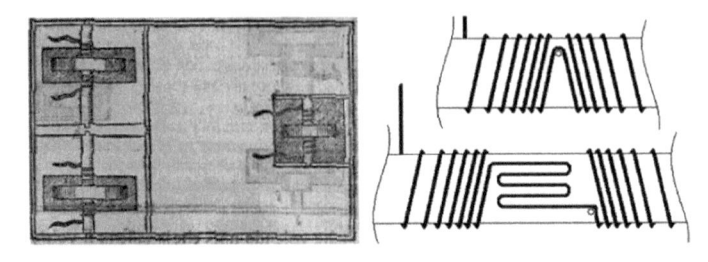

Fig. 4 Traction with independent axle shafts (*left*) and scheme of rope rolling to program the motion (*right*)

rope is rolled on one of the axle shafts in a different way from the other axle shaft, one of the two driving wheels will rotate in different way from the other one.

Moreover, it is also possible that, during the counterweight's motion, one of the wheel stops while the other rotates; this is obtained by wrapping a piece of the rope in an hank like shown in Fig. 4, on the right: during the time in which the hank unleashes, that axle shaft stops. It is also possible to obtain that one of the axle shafts rotates in the opposite sense with respect to the other one; this is simply obtained by rolling the rope on one axle in the opposite sense with respect to the other one.

Finally even a programming of the motion can be obtained by putting some knobs on the axle shaft (see Fig. 6); by means of which it is possible to modify the rolling of the rope, obtaining different laws of motion for each wheel.

This represents the first example of motion programming.

Some scholars (see e.g. [7]) have built models of carts moved by counterweight motors based on the works by Heron; they demonstrated, practically, the possibility of programming the motion.

It is also possible to obtain that one of the axle shafts rotates in the opposite sense respect the other one; this is simply obtained by rolling the rope on one axle in the opposite sense in respect to the other one. Finally, even a programming of the motion can be obtained by putting some knobs on the axle shaft as shown in Fig. 6; by means of these knobs it is possible to modify the rolling of the rope, as described before, in order to obtain different laws of motion for each wheel.

Some scholars (see e.g. [7]) have built models of carts moved by counterweight motors based on the works by Heron, demonstrating the possibility of programming the motion.

2.3 Siege Towers

These war machines were commonly termed in the Classic Age as "helepolis" (ἑλέπολις ≈ "taker of cities"). They have been widely described since antiquity by many authors, see. e.g. Diodorus Siculus (1st century B.C.) [2, 12], Publius Flavius Vegetius Renatus (IV–V century A.D.) [13], Cesare [14] and others and were commonly used till the Middle Ages. Pictorial reconstructions by the authors and his colleagues of a siege tower [1, 2, 11] were developed; they are based on several classics among which the most interesting can be found in [13–15].

These machines were certainly taller than the walls of the besieged town; therefore an average height of about 30 m can be considered, but much taller towers were also described. The base was rectangular or square, the length of the sides being roughly equal 1/5–1/3 of tower's height; the structure was generally tapered on the upper side. The front and perhaps also the sides were covered by metal plates [12] against projectiles thrown by the defenders; the "armour" was completed by a curtain of untanned and loose wet leathers that defended the tower from the incendiary projectiles. Some wheels were installed under the machine. The mass

Fig. 5 Scheme of he heletpolis' counterweight motor

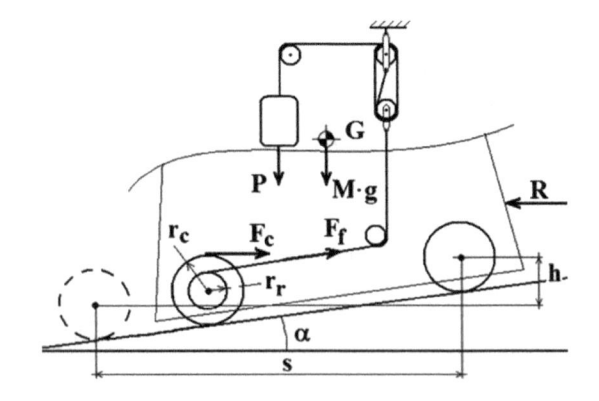

probably ranged between 30000 and 50000 kg. Hence it was certainly very difficult to move such big machines, even if the ground was prepared with a track made of wooden boards. In fact, to suppose that the helepolis were pushed or pulled by oxen or by a system of ropes and pulleys (the latter installed on poles rammed down at the base of the town walls) seems unrealistic. As for this aspect we can recall a piece by the Byzantine historian Procopius of Caesarea ($\sim$500–565 A.D.) on an unsuccessful siege of Rome from the Goths: Vitige, the king of the Goths (Wittigeis, ?—540 A.D.) used wooden siege towers pulled by oxen; the defenders easily killed the oxen and the towers became useless. Moreover, Julius Caesar (De Bello Gallico [14]) reports, during the siege of a town of the Gauls Atuatuci, the Gauls surrendered because they saw very big machines moving without any external mover.

Solution for potential internal motors were proposed [2, 11]; among those, in a study from the author [11] a counterweigh motor (like the one for the Hero's automata) seems to be the most effective and reliable. In Fig. 5 is reported a scheme of author's reconstruction of a possible counter-weigh motor for the propulsion of the helepolis. This hypothesis is also supported by a piece by Siculus [15].

2.4 Water Wheels

The end point of those ancient water turbines is the saw powered by a water wheel represented in the bas relief reported in the upper part of Fig. 6 together with an author's reconstruction. The bas relief was found on a cover of a sarcophagus at Hierapolis of Frygia (Asia Minor). It was dated by Professor Tullia Ritti (University of Naples "Federico II") to the 3rd Century A.D. So it does not strictly belong to Hellenistic period but represents an end point of the Hellenistic knowledge.

The device essentially consists of a double saw to cut marble powered by a water wheel. The saws are moved by the wheel shaft through a gear train and a crank and shaft mechanism. The latter, before this discovery, was commonly considered as a very later invention.

Fig. 6 Bas-relief representing the Hierapolis saw (Maier-Courtesy of prof. F. D'Andria, Missione Archeologica Italiana a Hierapolis di Frigia) and Pictorial Reconstruction of the Hierapolis Saw

2.5 Archimedes' Cannon

Although Archimedes doesn't strictly belong to the Hellenistic Age, his inventions and his thoughts belong to the same culture. At the end of the XV century, Leonardo Da Vinci drew a steam cannon that he ascribed to Archimedes and that, for a tribute to Archimedes, he called "architronito" (Thunder of Archimedes); the drawing is shown in Fig. 7. On the same folio is reported the working principle: a proper amount of water is put in the reservoir A, then the valve B_1 is opened and the water fills the tank C. Next the valve B_1 is shut and the valve B_2 is opened: the water flows in the chamber of the cannon and vaporizes. Through the pipe D, the pressure in the tank C is equalized to the one in the chamber of the cannon. The steam pressure throws the ball E outside the barrel.

The Greek origin of the device drawn by Leonardo da Vinci is demonstrated also by the units of measurement he reports that are ancient Greek units unknown in Italy in those times.

Before Leonardo Da Vinci, several authors described similar devices; among them Francesco Petrarca (Petrarch 1304–1374) that, in a minor work (De Remediis Utriusque Fortunae) describes a steam cannon about one century before Da Vinci.

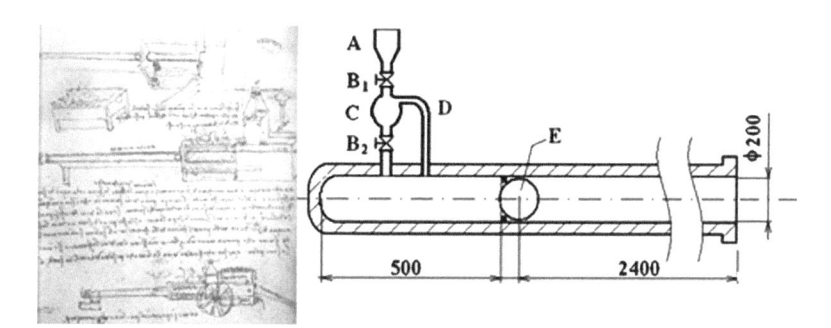

Fig. 7 Drawings by L. Da Vinci (Ms. B, f. 33 v) of the architronito and a scheme of the device based on the drawings and the brief description in Da Vinci's manuscript

The Greek historian Plutarchos (Vite parallele, vol. II, Pelopida e Marcello 14–15) tells that, during the siege of Syracuse, when the Romans saw something similar to a pole protruded from the walls ran away shouting : "Archimedes is going to throw something on us now". Now, no ancient throwing machine (such as onager, ballista or catapult) looks like a pole [2, 16–19]. Simms [20–22] cites a piece by Niccolò Tartaglia (Italian mathematician, about 1499–1557) where Valturius (Roberto Valturio, Italian engineer and literary man 1405–1475) "... States that ... there are many references to Archimedes having designed a device made from iron out of which he could shoot, against any army, very large and heavy stones with an accompanying loud report."

Finally, as it was already remarked by several investigators, no mention about burning mirrors was made by the historians of the Greek-Roman era but this legend appears only during the middle age.

The author supposed that such a cannon could have thrown hollow clay balls filled with an incendiary mixture called "Greek Fire". The possibility that the roman ships were burned by Archimedes by means of something like the famous "Greek fire" is also suggested by Simms [21], who reports that Galen (Aelius Galenus or Claudius Galenus or Galen of Pergamum 129–216) in his De Temperamentis says that "... Archimedes set on fire the enemy triremes by means of pureia (πυρεια)." Now, this word in ancient Greek indicates something used to light fire or can be translated as "brazier" but not as "burning mirror".

To evaluate the feasibility of such a device, the author and his colleague [23, 24] computed the pressure diagrams in the cannon and the ball range figures. The cannon dimension are reported in Fig. 7 while assumptions were made:

- internal volume: 0.111 m^3;
- minimum volume corresponding to the initial ball position: 0.035 m^3;
- mass of water introduced inside the cannon: 0.11 kg;
- mass of the projectile: 6 kg;
- cannon temperature in three cases: 430, 450, 470 °C;
- friction between projectile and barrel: 50 N
- heat transfer coefficient to the water: 10 kW/m^{2}K

Considering a tangential inflow of the water, the heat transfer surface was considered as the internal surface of the cannon and the breech.

The results obtained from the algorithm are summarized in Fig. 8.

2.6 Torsion Motors

According to several Authors [25, 26], the Greeks from Syracuse developed the first throwing machines using a torsion motor. This motor was much more advanced and powerful that the older ones based on the flexion of an elastic leaf (like in the bows) and consisted of a strong wooden square frame, reinforced by iron straps, divided into three separate sections. The central section was used to insert the shaft of the

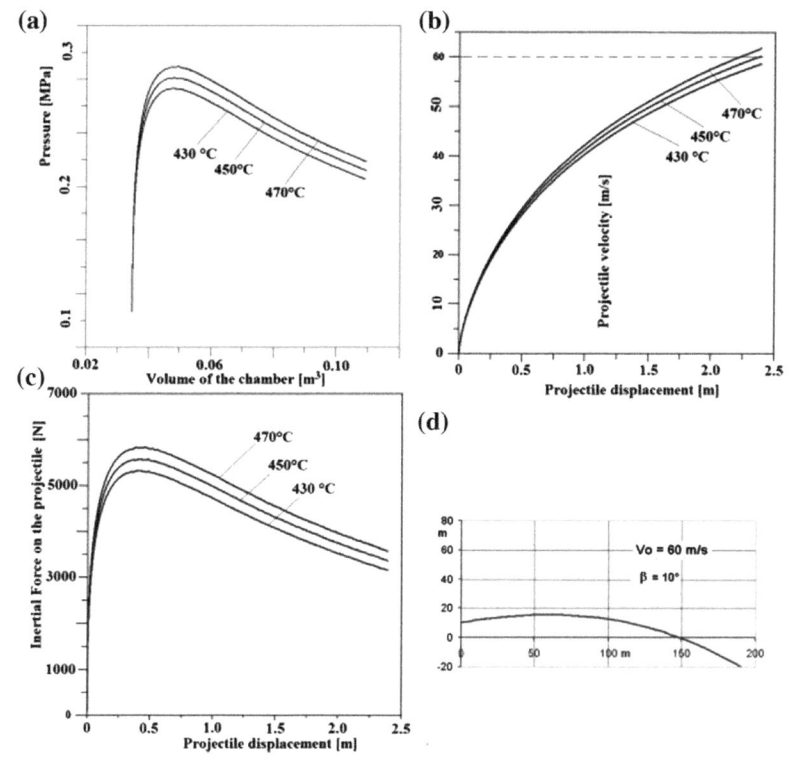

Fig. 8 Simulation results for the Archimedes' cannon: **a** the pressure volume diagram; **b** the projectile velocity versus its displacement along the barrel; **c** the inertial forces (in N) acting on the projectile along the barrel; **d** the projectile trajectory for a muzzle velocity of 60 m/s and a barrel elevation of 10°

weapons, while the sides were for the two coils made by a bundle of elastic fibers; the most widely used were women's hair [2, 27] being the natural fibers having the best mechanical properties. In Fig. 9 a motor of a Roman catapult is shown; on the left a remain found in Xantem, Germany is presented, in the middle an authors' pictorial exploded view [2, 27] and the bundle on the right.

The design of the Greek-Roman throwing machines was based on a module, i.e. the internal diameter of the *modiolus* (see Fig. 9). Probably the first ancient scientist who formulated a relationship between the weight of the projectile and the modulus diameter was Archimedes of Syracuse.

From Philon of Byzantium [16, 17] to Vitruvius [3], all throwing machines designer and theoreticians say that this relationship is:

$$D = 1.1 \cdot \sqrt[3]{100 \cdot m} \qquad (1)$$

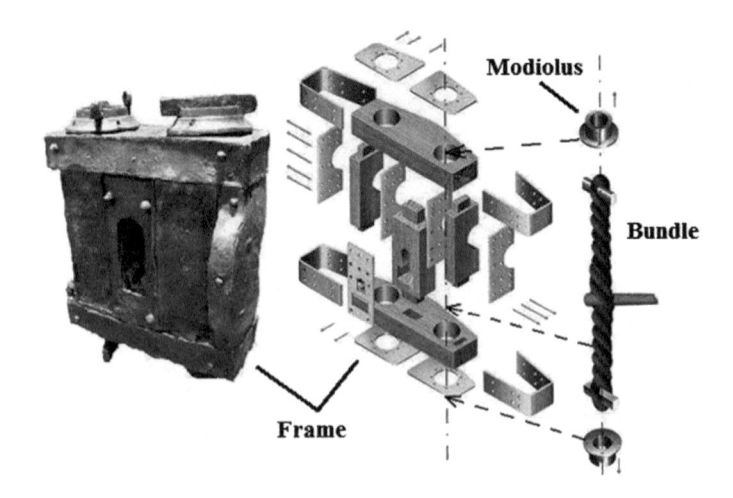

Fig. 9 Propulsor of a Roman catapult

where:
D is the diameter of the modiolus (hence of the hair bundle) in digits (1 digit ≈ 19.5 mm)
m is the mass of the projectile in mine (1 mina ≈ 431 grams).

Vitruvius [3] is very meticulous in giving the ratios between the diameter of the modiolus and all the other main dimensions of the machine; this clearly shows that ancient engineers designed their machines by adopting a modular design concept.

The model of the bundle of hairs representing the torsion motor was presented in [27] so just the main results are summarized. The elastic energy L stored in an hair bundle is:

$$L = E\pi l_0 \left\{ \frac{R^4\theta^2}{4 l_0^2} + R^2 - \frac{2 l_0^2}{3\theta^2}\left(\frac{R^2\theta^2}{l_0^2}+1\right)^{\frac{3}{2}} + \frac{2 l_0^2}{3\theta^2} \right\} \quad (2)$$

And the couple C exerted by the torsional motor is:

$$C = \frac{dL}{d\theta} = 2E\pi l_0 \left[\frac{R^4\theta}{2 l_0^2} - \frac{4 l_0^2}{3\theta^3} + \frac{4 l_0^2}{3\theta^3}\cdot\left(\frac{R^2\theta^2}{l_0^2}+1\right)^{3/2} - \frac{2R^2}{\theta}\cdot\left(\frac{R^2\theta^2}{l_0^2}+1\right)^{1/2} \right] \quad (3)$$

where (S.I. units):
E is Young's modulus of the hair yarns;
l_0 is half-length of the bundle;
R is the radius of the bundle (hence of the modiolus)
θ is the torsion of the bundle, given by the machine arms rotation and by the preload.

The L/D ratio between the length of the bundle and its diameter was decisive in obtaining the maximum energy from the bundle itself without exceeding the hair stress proportionality limit. It can be computed by [1]:

$$L/D = \sqrt{\frac{\theta^2}{\left(\frac{\sigma_e}{E'}+1\right)^2-1}} \cong \sqrt{\frac{\theta^2}{0,035}} \Rightarrow \theta_{max} \cong 10.7 \cdot L/D \ (deg) \qquad (4)$$

where σ'_e represents hair yarns stress proportional limit.

It was found that all torsion motor throwing machines were designed with that L/D ratio that permitted to achieve this stress proportionality limit at the maximum arms rotation for each kind of machine [27].

2.7 The Repeating Catapult

The invention of the repeating catapult is attributed to Dionysius of Alexandria, (III Century B.C.) and was described by Philon of Byzantium [16–18, 28, 29]. It can be considered as a concentration of the most advanced mechanical kinematic and automatic systems of the time, many of which are still widely used. A pictorial author's reconstruction [28], based on previous works and from the description by Philon [16] is shown in Fig. 10.

The device is fully automatic and consisted of a container holding within it a number of arrows, a cylinder feeding device and movement chain. According to Philon, the arrows were located in a vertical feeder F and were transferred one at a time into the firing groove by means of a rotating cylinder C, activated alternatively by a guided cam, in turn activated by a slide. One of the longer interior teeth T of the chain pulls the slide S which in turn pulled the cord, loading the coils of the motor. In the same time, an attached cam caused a 180° rotation in the direction of the cylinder, drawing an arrow from the loader and placing it in the channel in front of the rope. When the slide reached the rear of the weapon, the cog released it, while another one opened the release mechanisms. An instant later, upon completion of sprocket rotation, the same cog coupled with the slide from underneath, pulling in the opposite direction. Near the top of the weapon, the second device

Fig. 10 Views of the reconstruction of the repeating catapult

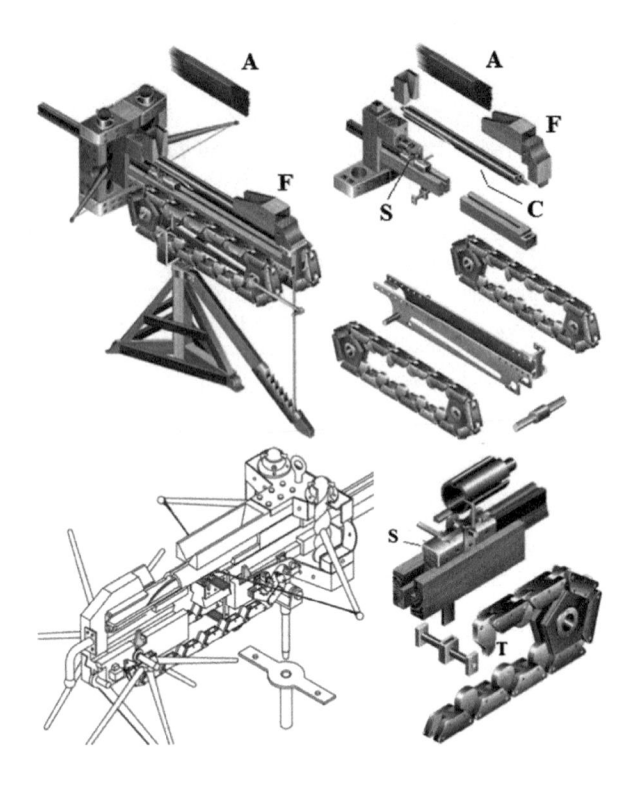

closed the hook after it had retrieved the cord, while the feeder cylinder picked up another arrow from the feeder. A half rotation in the sprocket and the cycle was repeated. It must be observed that our reconstruction [5], based on our translation of text by Philon, is really automatic; this because, as opposed to previous reconstructions [16, 30, 31], a simple rotation of the crank was sufficient to move the cylinder, the slide, the slide hooking mechanism and the trigger mechanism.

3 Conclusions

It is impossible to show an adequate number of examples of the Hellenistic knowledge in a few pages; so the author hopes that the few presented could be considered among the most interesting ones and can give an idea of how advanced the Hellenistic technical-scientific knowledge was. The Hellenistic knowledge was a basis for the later inventions made by Arab and European engineers and in some fields it was not improved till the 18th Century [1, 2, 32–36].

References

1. Rossi, C.: Some examples of the Hellenistic surprising knowledge: its possible origin from the east and its influence on later Arab and European engineers. Rivista Storica dell'Antichità **XLIV**, 1–84 (2014). ISSN 0300-340X
2. Rossi, C., Russo, F., Russo, F.: Ancient Engineers' Inventions. Springer, Precursors of the present (2009). ISBN 978-90-481-2252-3
3. Vitruvius, M.: De Architectura. Liber X
4. Russo, F., Rossi, C., Ceccarelli, M., Russo, F.: Devices for distance and time measurement at the time of the Roman empire. In: International Symposium on History of Machines and Mechanisms: Proceedings of Hmm 2008, pp. 101-114. Tainan, Taiwan, 10–14 Nov 2008
5. Baldi, B.: Di Herone Alessandrino de gli avtomati ouero machine se moventi libri due. Tradotti dal greco da B. Baldi. - Venezia 1602. http://echo.mpiwg-berlin.mpg.de/ECHOdocuView/ECHOzogiLib?url=/mpiwg/online/permanent/library/M5C8103Y/pageimg&pn=36&ws=2&mode=imagepath&start=1
6. Webb, B.: The first mobile robot? In: Proceedings of TIMR 99 Towards intelligent Mobile Robots (1999)
7. Sharkey, N.: The programmable robot of ancient Greece—new scientist issue 2611, pp. 32–35. 4 July 2007
8. Heronis opera quae supersunt omnia ed. W. Schmidt/L. Nix/H. Schöne/J. L. Heiberg Leipzig 1899 sqq. http://www.hs-augsburg.de/~harsch/graeca/Chronologia/S_post01/Heron/her_autp.html
9. Heron of Alessandria – Perì Automatopoietikès – http://www.hs-augsburg.de/~harsch/graeca/Chronologia/S_post01/Heron/her_autp.html
10. Baldi, B.: Di Herone Alessandrino de gli avtomati ouero machine se moventi libri due. Tradotti dal greco da B. Baldi. - Venezia 1602
11. Rossi, C., Pagano, S.: A study on possible motors for siege towers. J. Mech. Des. **133**, 1–8 (2011). ISSN 1050-0472
12. Siculus, D.: Bibliotheca historica, liber XX – from: Diodori Bibliotheca historica. Bekker et L. Dindorf; recogn. Fr. Vogel B. G. Teubner, 1985. Bibliotheca scriptorum graecorum et romanorum Teubneriana, Fac-sim. de la 3e éd., [s.l.] : [s.n.], Chapitre96 (1890). http://mercure.fltr.ucl.ac.be/Hodoi/concordances/diodore_20/lecture/default.htm
13. Publius Flavius Vegezius Renatus - Epitoma Rei Militaris, Liber IV, cap.XVII. http://www.thelatinlibrary.com/vegetius4.html
14. Cesare, G.G.: De Bello Gallico, Liber II, XXX et XXXI
15. Siculus, D.: Bibliotheca historica, liber XX – from: Diodori Bibliotheca historica. Bekker et L. Dindorf; recogn. Fr. Vogel B. G. Teubner, 1985. Bibliotheca scriptorum graecorum et romanorum Teubneriana, Fac-sim. de la 3e éd., s.l.. : s.n.. (1890)
16. Marsden, E.W.: Greek and Roman Artillery Historical Development. Oxford University Press II (1969)
17. Marsden, E.W.: Greek and Roman Artillery, pp. 106–184. Technical treatises, Oxford (1971)
18. Russo, F.: L'artiglieria delle legioni romane. Ist. Poligrafico e Zecca dello Stato (2004). ISBN 88-240-3444-6
19. Russo, F.: Tormenta Navalia. L'artiglieria navale romana, USSM Italian Navy, Roma (2007)
20. Simms, B.L.: Archimedes and the burning mirrors of syracuse. Technol. Cult. **18**(1), 1–24 (1977)
21. Simms, B.L.: Archimedes and the invention of artillery and gunpowder. Technol. Cult. **22**(1), 67–79 (1987)
22. Simms, B.L.: Galen on Archimedes: burning mirrors or burning pitch?. Technol. Cult. **32**(1), 91–96 (1991)
23. Rossi, C.: Archimedes cannon against the Roman fleet ? In: Plenary Lecture at the International Conference of "The Genius of Archimedes", pp. 113–131. Syracuse, Italy, 8–10 June 2010

24. Rossi C., Unich A. (2013) A Study on Possible Archimede's Cannon. Rivista Storica dell'Antichità. Vol. XLIII. ISSN 0300-340X
25. Chondros, T.G.: The development of machine design as a science from classical times to modern era HMM 2008. In: International Symposium on History of Machines and Mechanisms. Proceedings Published by the Springer, Tainan, Taiwan, Netherland, 11-14, Nov 2008. ISBN 987-1-4020-9484-2 (Print) 978-1-4020-9485-9 (Online)
26. Chondros, T.G.: Archimedes (287–212 BC) history of mechanism and machine science 1, Distinguished figures in mechanism and machine science, Their contributions and legacies, Part 1. Ceccarelli, M. (Ed.), University of Cassino, Italy. Springer, The Netherlands (2007). ISSBN 978-1-4020-6365-7
27. Rossi, C.: Ancient throwing machines: a method to compute their performances. Mech. Mach. Theory **51**, 1–13 (2012). ISSN 0094-114X
28. Rossi, C., Messina, A., Savino, S., Reina, G.: Performance of Greek-Roman artillery. Arms Armour, J. R. Armouries **12**(1), 66–88 (2015). ISSN 1741-6124, online ISSN 1749-6268
29. Rossi, C., Russo, F.: A reconstruction of the Greek-Roman repeating catapult. Mech. Mach. Theory **45**(1), 36–45 (2010). ISSN 0094-114X
30. Shramm, E.: Die antiken Geschütze der Saalburg. Saalburg Museum, Reprint, Bad Homburg (1980)
31. Soedel, V., Foley, V.: Ancient catapults. Sci. Am. (1979)
32. Rossi, C.: Guest editorial "On designs by ancient engineers". J. Mech. Des. Trans. ASME **135** (6), art. no. 060301 (2013). doi:10.1115/1.4024
33. Lazos, C.: Schematics of steam cannon built by I. Sakas; from "Archimedes: The Ingenious Engineer", p. 183. Athens (1995) (in Greek). http://www.cs.drexel.edu/~crorres/bbc_archive/steam_cannon_schematic.jpg
34. Archimedes's Steam Cannon (M.I.T. website). http://web.mit.edu/2.009/www/experiments/steamCannon/ArchimedesSteamCannon.html
35. Messina, A., Rossi, C.: Mechanical behavior and performance of the onager. J. Mech. Des. **137**(3), 034501 (2015) (5 pages); Paper No: MD-14–1276. doi:10.1115/1.4029319
36. Chondros, T.G., Milidonis, K., Paipetis, S., Rossi C.: The trojan horse reconstruction. Mech. Mach. Theory **90**, 261–282 (2015). ISSN 0094-114X

Mechanical Engineer DING Gongchen

Zhang Baichun and Liu Yexin

Abstract DING Gongchen (丁拱辰, 1800–1875), mechanical engineer, was one of the first people in China who systematically examined the application and the structure of Western firearms, and tried to study and produce firearms in China. He was also the first man in China constructing the models of steam engine, locomotive and steam boat, and wrote the first book on those machines and weapons. He has made great contributions to the development of mechanical engineering in the mid-19th century China.

Keywords DING Gongchen · Mechanical engineering · Firearms · China

1 Introduction

DING Gongchen (Fig. 1), alias Jun Zhen (君轸), styled himself Shu Yuan (淑原), with an assumed name of Xing Nan (星南), was born in a Hui family (one of the minority ethnic groups in China) in Jinjiang Xian, Quanzhoufu, Fujian Province in 1800. His ancestor was an Arab nobleman named Sayyid Ajjal Shams al-Din Omar (赛典赤·赡思丁) who became an official in Yuan China. Ding Zongbi (丁宗壁), DING Gongchen' father, was a small business trader. Due to destitution, DING Gongchen was forced to drop out of school at the age of eleven. But while working, he insisted on reading, sometimes studying art of war and eventually became one of the Imperial College Students. In 1817, he was engaged in business complying with his father's wish, but at the same time he kept reading. He became also interested in astronomical observation, designing instruments himself used for measuring the shadow of the sun and observing stars [1].

Z. Baichun (✉) · L. Yexin (✉)
Institute for the History of Natural Sciences, CAS, Beijing, China
e-mail: zhang-office@ihns.ac.cn

L. Yexin
e-mail: liuyexin@ihns.ac.cn

Fig. 1 DING Gongchen

In 1831, DING Gongchen made his living overseas in foreign merchant ships. He had been to places such as Philippines, Luzon and Arab countries, etc., during which he managed to get acquainted with those who were good at mathematics and those who were specialized in device-constructing, and to discuss with them. DING Gongchen carefully examined the ship structures and different types of guns when he was abroad. Influenced by what he had constantly seen and heard, he tried his utmost to study. After coming back to the motherland in 1840, DING Gongchen was deeply worried by what had happened to his country, which was desolated by opium smuggling and British invasion. Holding his firm patriotic ambition, DING Gongchen managed to improve the Chinese gunnery technology, and endeavored to popularize his achievements by spending all his money in engraving his book *Yan Pao Tu Shuo* (演炮图说, *The Illustrations and Descriptions of Guns-Operating*). In 1850, DING Gongchen was invited to Guilin to make weapons. The next spring, he was dispatched to Guangxi province to supervise gun-building used for suppressing the Taiping Rebellion. Soon after that, he was sent to Fujian province to train local militia. Later, he went abroad to engage in business again. In 1861, LI Hongzhang (李鸿章) recommended DING Gongchen to the Qing Court and asked him to assist with Western weapons management in the army. Appreciating his talents, LI Hongzhang had petitioned the Court to grant DING Gongchen the "feathered cap of the fifth rank", but he did not take up this post. It was said that he passed away in 1875.

2 Study and Design Firearms

After coming back to China in 1840, DING Gongchen, basing on his research on foreign guns, proposed that China should adopt the operational table of gun-aiming (火炮加表). Basing on the tests results, he had produced an illustrated book entitled *Yan Pao Cha Gao Tu Shuo* (演炮差高图说), which improved the hitting probability of guns. LIANG Baochang (梁宝常), the Guangdong Grand Coordinators, had witnessed in Guangzhou outskirt the success of this method, and reported the result to the Court who later granted DING Gongchen the "feathered cap of the sixth rank". In order to defend invasions, DING Gongchen compiled all the materials relating to Western weapons, and by further summarizing his own research results, he published the finalized *Yan Pao Tu Shuo* in 1841. This book mainly demonstrated the methods of gun-firing, gun-making, gun emplacements construction, and powder formulation, etc. In 1842 spring, this book was highly valued by QI Gong, Governor-General of Guangdong and Guangxi, and YI Shan, Jingni General (靖逆将军), who ordered XI Laben (西拉本) to help impart technology with DING Gongchen in the army. In the same year, the Qing Court intended to popularize the methods described in *Yan Pao Tu Shuo* and later DING Gongchen further edited the book into a four-volume *Yan Pao Tu Shuo Ji Yao* (演炮图说辑要, *The Important Collection of the Illustrations and Descriptions of Gun-operating*) published in the fall of 1843.

DING Gongchen's life was marked with many achievements on weapon design. The pulley-device (滑车绞架) he constructed could adjust the position of barrel assembly and change the angle of firing conveniently and easily. By using the pulley-device, only a few people, instead of dozens of them as used to be, were required to move and adjust the guns of thousands of Jins. Therefore, this design was applied in land fort barbettes and in war ships in Guangdong. At that time, either copper or iron was used to cast guns. As both of the quality and the cost of copper-casted guns were higher than that of iron-casted guns, DING Gongchen managed to improve the performance of iron-casted guns by drawing lessons from the structure of Western guns. After many investigations and trials, DING Gongchen found that by mixing two kinds of irons produced in Guangdong in a ratio of three to seven, a different type of iron in a better quality could be casted. By further adopting clay mould and setting the sprue at gun muzzle, DING Gongchen eventually built a high-quality iron gun. He noticed that the shells made in the West by the clay mould process tended to be left with striae on the surface, which would make its surface uneven and affect the range of fire. Therefore, DING Gongchen adopted the lost wax process to cast shells which would ensure a smooth and stria-free surface. He also advocated making two different kinds of shells, solid and hollow. The advantages of the latter were light in weight and had a longer range of fire. DING Gongchen had made hundreds of various guns when he was in Guilin cooperating with DING Shoucun (丁守存, 1812–1885). He also made other machines such as rocket, jingalls, shotguns, gunpowder, etc.

3 Trial-Producing of Engine Machine, Locomotive and Steam Boat

According to the record of *Xi Yang Huo Lun Che, Huo Lun Chuan Tu Shuo* (*Illustrations of Western Steam locomotives and Steam boats*, 西洋火轮车、火轮船图说) compiled in his works of *Yan Pao Tu Shuo Ji Yao*, DING Gongchen had once seen a small type of steam locomotive in person. Knowing something about machine-building, he invited a good carpenter to calculate the size for him and he constructed a model of steam locomotive himself [2]. Its body was 1 chi and 9 cuns' long (632.7 mm), 6 cuns' wide (199.8 mm), with 30 jins in weight (15 kg), equipped with a copper-made vertical reciprocating steam engine (Figs. 2 and 3).

Basing on this steam engine, DING Gongchen had also constructed a model of steam boat with a length of 4 chis and 2 cuns' (139.86 mm), moving fast in the water. But due to its small size and light weight, it could not go too far. DING Gongchen later had drafted a picture of steam boat, but it failed to be built into something practical because steam boat-building machine was not available by then in China (Fig. 4). Even though, the work that DING Gongchen had done represented a new era for the development of Chinese mechanical engineering technology.

In 1850, basing on the new-style British rockets, DING Gongchen and DING Shoucun in Guilin, Guangxi Province constructed successfully a different kind of rocket with the range of firing of more than 200 Zhangs (660 m). At the bottom of the rocket there were five holes (five jet nozzles) jetting out flames, the smoke of which would soon permeate into everywhere. It performed well on setting fire on enemies' camps and charging and destroying enemies' positions. Under such attack anyone would certainly die, which was as fierce as guns. This is the beginning of research and design of rockets in Chinese history [3].

Fig. 2 Steam locomotive drawn by DING Gongchen

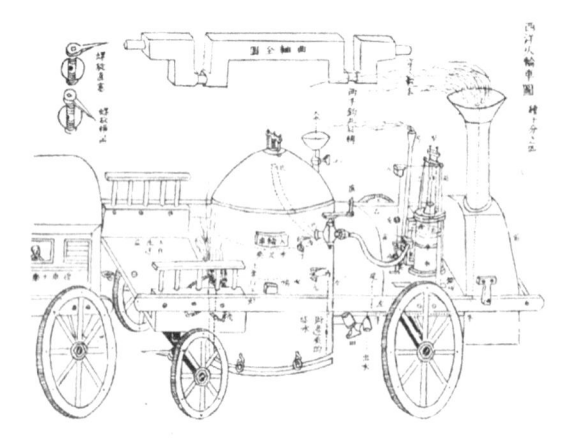

Fig. 3 Mechanical drawings of steam locomotive

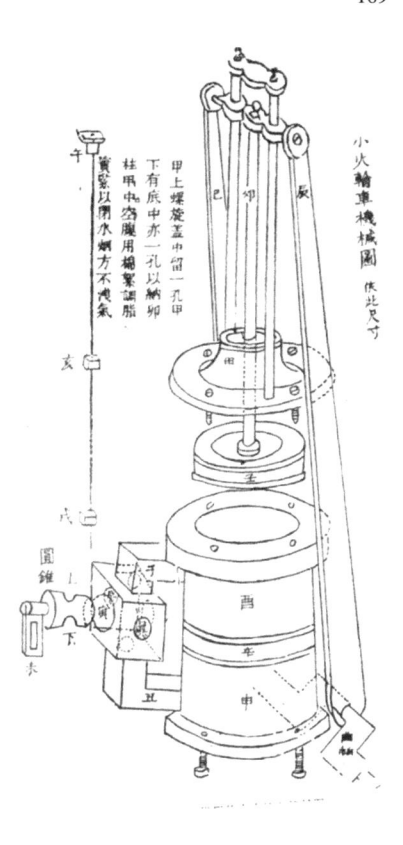

Fig. 4 Steam boat drawn by DING Gongchen

4 Conclusions

(1) DING Gongchen was the first man in China who illustrated and explained to the Chinese people the knowledge of steam engine in the 19th century. He successfully constructed the models of steam engine, steam locomotive and steam boat, which is significant in the early modernization of machine-building in China's history. It was not until 20 years later that Chinese people eventually patterned the steam locomotive and steam boat of practical use in success.
(2) DING Gongchen, also one of the main pioneers in firearm-constructing in early modern China, had introduced and improved several crucial weapons in his lifetime during which China had suffered a lot due to foreign powers' invasion. His achievements in guns had enhanced the combat capability of the army and had won him the appreciation of both the Qing Court and those officials with heavy responsibility.
(3) Through a detailed case study on DING Gongchen, it shows clearly that even in one of the most miserable periods in Chinese history, there still were few Chinese people who kept, introducing and practice new knowledge from the West to equip China. DING Gongchen, undoubtedly, was one of the earliest and the most prominent figures treating those new foreign technologies with an open mind.

References

1. Lingen, X., Gongchen, D.: Scientists in the Pre-Modern China, pp. 59–67. Shanghai People's Publishing House, Shanghai (1988) [in Chinese]
2. Gongchen, D: Xi Yang Huo Lun Che, Huo Lun Chuan Tu Shuo. In: Yan Pao Tu Shuo Ji Yao, vol. 4, pp.13–19 (1843) [in Chinese]
3. Jixing, P.A.N.: History of Chinese Rocket Technology, p. 77. Science Press, Beijing (1987) [in Chinese]

Lewis Mumford Revisited

Teun Koetsier

Abstract In this paper we use Mumford's thesis, that the clock and not the steam engine is the key to the modern industrial age, as a starting point to make some remarks on the role of the clock in Western culture. We briefly discuss the mechanical clock in the West, its invention and reception. The relation between the clock and the Industrial Revolution is threefold. Its role in the synchronization of human activity is of crucial importance. In different ways the clock and the Scientific Revolution are connected: several scientists were involved in improving clocks and clock makers and the makers of scientific instruments are two of a kind. Moreover clock technology was crucial in the course of the Industrial Revolution.

Keywords Lewis Mumford · Industrial revolution · Clock · Steam engine · Scientific revolution

1 Introduction

In 1934 Lewis Mumford argued that the mechanical clock, not the steam engine, is the key to the modern industrial age [1]. This is a remarkable thesis, because in many histories of technology the clock plays no role whatsoever or only a small role. The Industrial Revolution is a complex social development. It gains momentum in the 18th century with the mechanization of spinning and weaving. It starts with the flying shuttle, spinning machines and improved looms. At the end of the 18th century Watt's innovations turn the steam engine into the most promising source of power. Slowly the steam engines would take over and dominate the 19th century. When we look at the machines that everybody associates with the Industrial Revolution the clock only seems to play a marginal role.

T. Koetsier (✉)
Department of Mathematics, Faculty of Science, VU University,
Amsterdam, Netherlands
e-mail: t.koetsier@vu.nl

The second major cultural transition that brought about the modern age is the Scientific Revolution. In the literature about this revolution the clock is usually hardly mentioned. A representative example is the classic text on the background of the Scientific Revolution by Dijksterhuis, *The Mechanization of the World Picture* [2] in which the clock is basically absent.

Mumford was a philosopher in search of the essence of things. In *Technics and Civilization* he argues that the essence of what happened is not in the weaving machines or the steam engines. They represent the surface. Beneath the surface two waves of industrial revolution can be distinguished. The first one started in the 10th century when inside Western Culture discipline and order gained importance. The clock brought about regimentation and without regimentation mechanization alone could not come to dominate culture. Clocks do not only measure time, they synchronize the acts of man. The second wave coincides with what we usually call the Industrial Revolution. In this second wave the synchronization of human activity reaches unprecedented heights. The new machinery and its use illustrate this in many ways ([1], p. 84). There is another aspect introduced by Mumford. He argues that the 19th century machine is a combination of two things, the clock and the gun. The clock represents order and the gun represents power. The 19th century machines are often the combination of these two elements.

In the present paper we will take Mumford's thesis seriously. Let us investigate the thesis that the clock is the key to the modern industrial era.

2 The Invention of the Mechanical Clock

By the ninth century the Benedictine form of monastic life, defined by Saint Benedict of Nursia (c.480–543), had become the standard throughout Western Europe. The *Rule* of Benedict consists of 73 chapters. Its wisdom is spiritual and administrative. One-tenth of the chapters, however, outline how the monastery should be managed.[1] This *Rule* has been extremely influential. The division of the day in parts had the full attention of the Benedictine monks. Also Bede's *Temporum Ratione* shows clearly that the division of the day was a matter of interest.

All over Western Europe the monasteries were ready for the mechanical clock. It seems to have been invented between 1200 and 1300. The crucial invention was the verge-and-foliot escapement in combination with a crown wheel driven by a falling weight. This type of escapement was used for several hundred years. The foliot is often a horizontal rod with two weights attached to it. The escapement transforms the continuous accelerated rotation caused by a falling weight into a uniform rotary motion consisting of a sequence of escapes separated by stops (Fig. 1).

---

[1]https://en.wikipedia.org/wiki/Benedict_of_Nursia#Rule_of_St._Benedict.

Fig. 1 Verge and foliot escapement

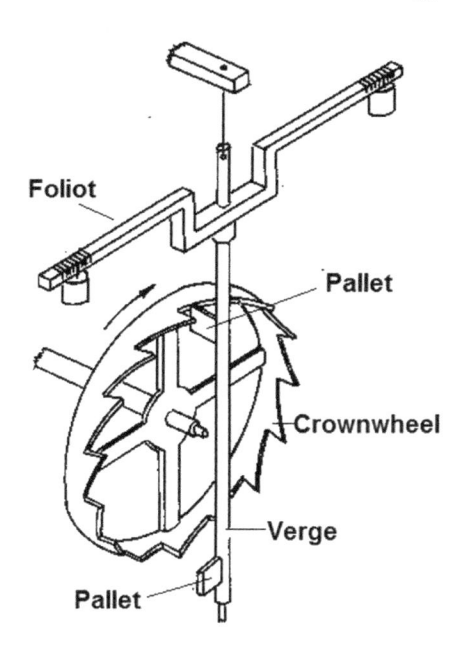

Soon the clocks became quite complex and they were doing more than merely show the time in a simple way. For example, between 1348 and 1364 in Padova the Venetian clock-maker Giovanni Dondi dell' Orologio built a very complex astronomical clock. This *Astrarium* had seven faces showing the positions of the sun, the moon and the five planets.

In Fig. 2 from De Dondi's text the foliot has the form of a horizontal crown. It is fixed to a vertical rod, the verge, which has two pallets attached to it. The verge and foliot is made to rotate to and fro by means of a vertical crown wheel that is via a train of wheels ultimately driven by a weight. A tooth of the crown wheel pushes against one of the pallets and makes the verge rotate. This motion brings the other pallet in front of another tooth of the crown wheel. It stops the tooth and is then pushed backwards until the other pallet stops the wheel again.

Although the first clocks were not very accurate they captured the imagination. The clocks seemed alive. The escapement is the source of the characteristic ticking sound of mechanical clocks and it resembles the heartbeat of a living creature. Lynn White wrote: "Suddenly, towards the middle of the fourteenth century, the mechanical clock seized the **imagination** of our ancestors. Something of the civic pride which earlier had expended itself in cathedral-building was diverted to the construction of astronomical clocks of astounding intricacy and elaboration" [3].

Fig. 2 The *Astrarium*: modern tracing of an illustration in the *Tractatus astrarii*. The drawing does not show the complex upper section of the clock with its wheels and dials for the Sun, the Moon and the planets (*Source* https://en.wikipedia.org/wiki/Giovanni_Dondi_dell%27Orologio)

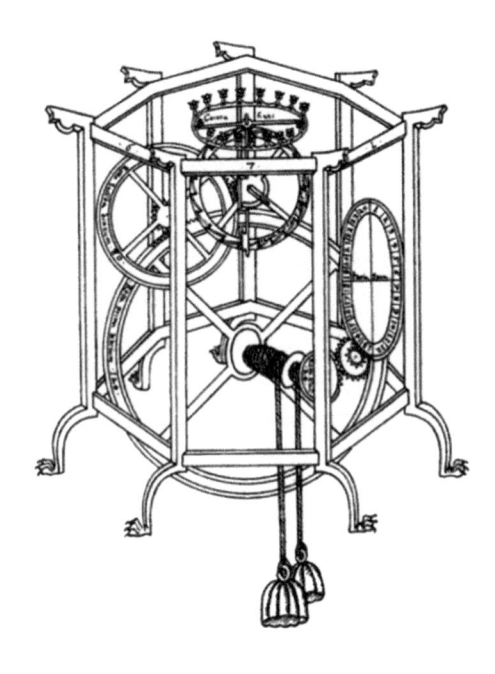

3 The Reception of the Clock in the West and the East

The early clocks were not very accurate and until the 19th century they had to be synchronized by means of sun dials. Yet they slowly gained popularity. In the cities, on towers, clocks started appearing and in the 15th century clocks entered the homes of our wealthy ancestors.

In the 16th century the first watches were made. An example of an early watch is the Nuremberg egg: a small spring-driven clock made to be worn around the neck, made in Nuremberg.

In the 17th century Christiaan Huygens proved that a pendulum swinging between two the cycloid guides would have a regular swing independent of the amplitude. He replaced the foliot by a pendulum and soon the Golden Age of pendulum clocks started. In 1675 Huygens invented the balance spring. Robert Hook made the same invention but only applied it after having heard about Huygens' work. The spring made much more accurate watches possible. Here scientists are influencing clock making and less than a century later clock making would decisively influence the development of the machines of the Industrial Revolution, as we will see (Fig. 3).

There were setbacks. In 1665 the plague hit London. Many left the capital while many others died. This was followed in 1666 by the Great Fire of London that burned for over 3 days and destroyed most of the city, including more than half of

Fig. 3 *Left* Early Nuremberg egg from circa 1510 (https://en.wikipedia.org/wiki/Nuremberg_eggs). *Right* Dutch clock from about 1740 (Courtesy of the Zaan Time Museum)

the approximately 160 clock makers' workshops.[2] Of course the clock makers' trade recovered (Figs. 4, 5 and 6).

It is interesting to compare the development in the West with the developments in the East. Between 800 and 1400 China Chinese society reached a level of civilization comparable to that of the Roman Empire at its peak. The West lagged behind. Yet the West was slowly recovering from the dark cultural and economic deterioration that had followed the decline of the Roman Empire. The West was on its way back to the top and the East did not keep up with it. The Jesuits introduced the mechanical clocks in China. The Chinese were interested but did not realize what the clocks represented. They viewed themselves as culturally superior and were not inclined to leave this position. They were right at the time, but this would change.

Time was viewed and experienced differently in the West and the East. In China officials came to ceremonies hours too early in order not to be late. If all time, also

---

[2]http://www.antiquesandfineart.com/articles/article.cfm?request=437.

Fig. 4 De Vaucanson's automatic flute player, the digesting duck and the Tambourine Player (1738–1739)

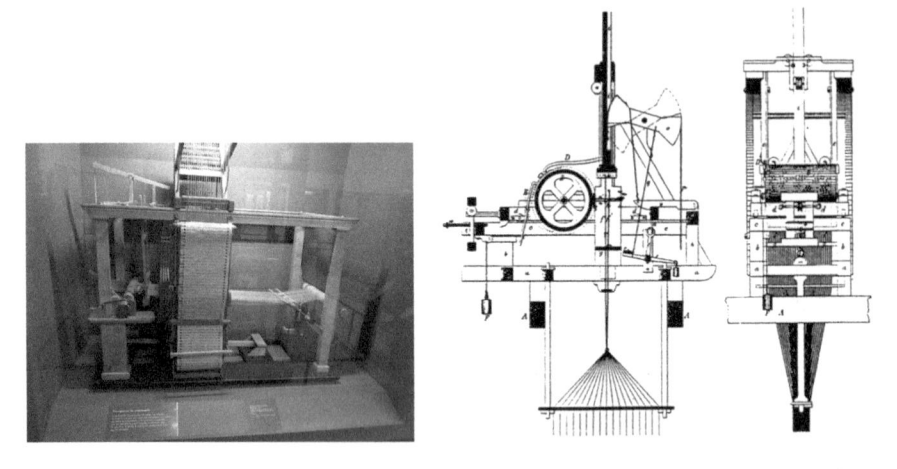

Fig. 5 Bouchon's 1725 loom in the *Musée des Arts et Métiers* (*left*) and a drawing of De Vaucanson's 1745 loom (*right*)

your time, belongs to the Emperor, why bother? In the West the clock represented and introduced a new awareness of time. This was economically based but there was also a religious impulse: Memento mori, don't waste your time.

Fig. 6 Providence and Worcester Railroad 1853. Fourteen killed, caused by a faulty watch

The Protestant work ethic is a concept coined by Max Weber in his book *The Protestant Ethic and the Spirit of Capitalism* [4]. It expresses that in Protestantism the focus is on hard work and modesty in contrast to the emphasis on ceremony in the Catholic tradition. David Landes, who has described the impact of the watch in *Revolution in Time* (1983), has pointed out that the majority of clockmakers were Protestants. Landes: "the consumption of timepieces may well be the best proxy measure of modernization, better even than energy consumption per capita, which varies significantly with the relative cost of fuel, climatic requirements, and product mix." The use of watches shows "a whole bundle of new work and life requirements and the inculcation of the values and attitudes that make the system go".

With the Industrial Revolution the role of this awareness of time is very clear. Working in the factory required a new discipline. People coming from the village were used to control their own time. When to work, when to do nothing, when to visit the toilet, etc. Coming late or not at all were in the beginning frequently occurring. With people working sometimes 14 h per day this is not surprising. Sometimes the workers were fetched from their homes. There were fines and rewards. A clock was not an unusual reward. As a result the need for clocks was growing. At the end of the 18th century Great Britain produced 150.000–200.000 watches per year, more than half of the total European production. In 1790 Great Britain had a population of 8 million. That is one watch per 40 inhabitants. At the time there were 1000 steam engines in Britain: one steam engine per 8000 inhabitants. Steam engines were bigger but watches were much more numerous.

Between Prescot and Liverpool—8 miles—the landscape was covered with the houses of spring makers, wheel cutters, chain makers, dial-makers, watch case makers etc. A nice and early example of division of labor ([5], p. 231).

4 Understanding the World by Means of the Clock

In the course of time in the West the clock caught the imagination in other ways. The clock turned out to be a thing that helped to understand the functioning of the world. The fact that ticking of a clock is like a heartbeat, suggests that living beings might be similar to clocks in certain ways. This idea led the philosopher Descartes. He argued that the nature of the mind (a thinking thing) is completely different from that of the body (an extended, non-thinking thing). This is called the mind-body dualism. Physical and biological phenomena ought to be explained solely in mechanistic terms, in terms of matter and its motion. Here the clock is the metaphor. Descartes wrote in his *Treatise on Man* on the movement of the body: "I should like you to consider that these functions (including passion, memory, and imagination) follow from the mere arrangement of the machine's organs every bit as naturally as the movements of a clock or other automaton follow from the arrangement of its counter-weights and wheels" ([6], p.169). Descartes views were influential and definitely contributed to the Scientific Revolution.

Living beings were compared to clocks and so was the universe, a metaphor definitely suggested by the clocks combined with planetaria. Kepler wrote "My aim is to show that the heavenly machine is not a kind of divine, live being, but a kind of clockwork, insofar as nearly all the manifold motions are caused by a most simple, magnetic, and material force [vis magnetica corporalis], just as all motions of the clock are caused by a simple weight. And I also show how these physical causes are to be given numerical and geometrical expression" ([7], p. 345).[3]

The clock metaphor definitely played an important role in the Scientific Revolution.

5 The Clockmakers as the Specialists in the Art of the Transformation of Motion

When clocks became more complex the clockmakers developed into true specialists in the art of the transformation of motion. Moreover the clockmakers not only built clocks. They built planetaria, musical instruments and automata. A nice example is the Frenchman Jacques de Vaucanson (1709–1782). De Vaucanson wanted to become a clockmaker but specialized in the design and building of automata. His automatic flute player, the digesting duck and the Tambourine Player became very well known.

The duck could drink water, digest grain, and defecate. It could flap its wings which contained each allegedly over 400 moving parts. Actually there was no real digestion. The food entered one reservoir and excreted a mixture of bread crumbs

---

[3]Copied from http://galileo.phys.virginia.edu/classes/109N/1995/lectures/kepler.html.

and green dye from another reservoir inside the pedestal of the duck. Yet it was an impressive mechanism.

The close links between clock and automata constructors, on the one hand, and the textile machines of the Industrial Revolution are clear from the following example. In 1725 Basile Bouchon designed and built a semi-automatic loom based on a tape of perforated paper, similar to the piano roll applied at the end of the 19th century. In 1745 the automata builder De Vaucanson designed and built the first fully automatic loom.

De Vaucanson made his invention too early. When half a century later Joseph Marie Charles *dit* (called or nicknamed) Jacquard (1752–1834) saw the remains of De Vaucanson's loom in the *Musée des Arts et Métiers*, he realized its potential. With a few refinements the Jacquard-loom became a huge success.

The links between the clockmakers and the Industrial Revolution are even clearer in England. In 1768 Richard Arkwright built his water frame, a second generation spinning machine driven by water. Arkwright called the gear wheels in his water frame the 'clockwork' and hundreds of clockmakers made the wheels for the 150 spinning factories that existed at the end of the 18th century. These gear wheels were made on the same machines that were used for the production of gear wheels for clocks.

Without exaggeration we can say that the early engineering machine tools evolved from the lathes and wheel-cutting engines of the clock- and watchmaking trade. This has been already been pointed out by Robert Willis in his Great Exhibition Lecture of 1852.

Not by accident were Watt and Smeaton instrument makers. There is an interesting letter written in 1791 from John Rennie to Matthew Boulton in which he says that the cotton trade had deprived London of many of the best clock makers and many of the mathematical instrument makers so that they can scarcely be had to do their ordinary business. (Musson and Robinson [8], p. 438)!

6 Further Developments

Already in 1748 Benjamin Franklin wrote "Remember that time is money", but it took some time before the businessmen, those involved in transportation and the officials of the postal services started really appreciating promptness, regularity and speed. It started in the English factories where workers could no longer determine themselves when to start to work and when to stop.

A worker at Horatio Allen's Novelty Works in New York in the 1830s wrote: "In the first place the bells are rung at 4½ in the morning that is to get up at 7 min before five they ring a second time and then we have to go into the shop and commence work. At seven we wash and go to breakfast and come back a quarter before eight. At half past twelve to dinner and back quarter past one and to tea at seven when I felt so tired I have been to bed soon after".[4]

Soon the growing complexity of the economy made speedy delivery of news increasingly important and the news travelled by mail. In the US, for example, in 1828 the Postmaster General was firmly committed to speeding up the delivery of mail (Stephens [9], p. 61). The economies were moving faster and faster and investors, politicians, merchants and businessmen became more and more aware of the importance of time.

With the rise of the railroads mail started travelling by train. The railroads represent an area where watches became extremely important. Before the telegraph became common in the 1860s watches and time tables determined the functioning of the railroads. Faulty watches could cause accidents.

In the course of the 19th century time measurement and industrial innovation are intertwined in remarkable ways. A lovely example concerns the famous race horse Lexington. In 1855 it set a world speed record in a four-mile race: 7 min and 19 3/4 s. The record stood for nearly 20 years. Allegedly Lexington's record stimulated the American Watch Company in Waltham, Massachusetts to build the world's first mass-produced stopwatch.[5]

Less than half a century later the stopwatch would become the symbol of Frederick Winslow Taylor's extremely influential *scientific management*. Taylor's assistants timed workers' motions in order to determine the most efficient way of doing a certain task. Taylorism permeated the workplace all over the industrialized world (Fig. 7).

Mumford's thesis was that the clock and not the steam engine is the key to the modern age. Obviously the role of the clock in the development towards and in the

Fig. 7 *Left* The legendary Lexington (1850–1875) was the leading sire in North America 16 times. *Right* Chronodrometer, or improved horse timing watch, American Watch Company, Waltham, Massachusetts, about 1859 (*Source* National Museum of American History. See: http://www.americanhistory.si.edu/ontime/mechanizing/index.html)

---

[4]Stephens [9], p. 73.
[5]Stephens [9], p. 68.

Industrial Age has been considerable. Moreover, if we are forced to choose between the two, the clock and the steam engine, the fact that the steam engine has become relatively rare while a life without clocks is in our industrial world absolutely unimaginable, seems to give the clock a clear head start.

References

1. Mumford, L.: Technics and Civilization (1934)
2. Dijksterhuis, E.J.: The Mechanization of the World Picture (1961)
3. Lynn, W.: Medieval Technology and Social Change. Oxford (1962)
4. Weber, M.: The Protestant Ethic and the Spirit of Capitalism (1905)
5. Landes, D.S.: Revolution in Time: Clocks and the Making of the Modern World. Harvard (1983)
6. Gaukroger, S. (ed.): Descartes, The World and Other Writings. Cambridge (2004)
7. Koestler, A.: The Sleepwalkers. London (1989)
8. Musson, A.E., Robinson, E.: Science and Technology in the Industrial Revolution. Manchester (1969)
9. Stephens, C.E.: On Time, How America Has Learned to Live by the Clock. Smithsonian Institution (2002)

Analysis of Structure, Kinematic and 3D Modeling of Ferguson's Mechanisms

V. Tarabarin and A. Kozov

Abstract It is known that the mechanisms invented by Scottish astronomer and engineer James Ferguson in 18th century, are of an interest for modern mechanics. While developing mechanisms, which simulate motion of celestial objects (so-called Planetariums and Lunariums) he discovered some unusual properties of planetary gear mechanisms called "Ferguson's Paradox". There are several models in the collection of Bauman Moscow State Technical University (BMSTU) department "Theory of Mechanisms and Machines" (TMM). This paper focuses on a structure, kinematics and a development of 3D models of mechanisms. The article will be of interest for technical institute professors and students, as well as for all readers who are interested in the problems of the Mechanical science and its history.

Keywords History of technology · Ferguson's paradox · Planetary mechanisms · Degree of freedom · Kinematic research · Velocity ratio · 3D mechanical models

1 Introduction

James Ferguson's life and works are described fairly enough in his biography [1, 2], which has been published more than once [3]. Ferguson was born on 25th April 1710 near Keith in Banffshire in Scotland in a large peasant family. The family lived by cultivating several acres of rented land. They had no money to give good education to their children, so James attended Keith's school for only 3 months. He learned to read himself. James spent his childhood and youth in Scotland, earning money by painting portraits. He married Isabel Wilson on 31st May 1739, who bore him four children.

V. Tarabarin · A. Kozov (✉)
Bauman Moscow State Technical University, Moscow, Russia
e-mail: alexey.kozov@gmail.com

V. Tarabarin
e-mail: vtarabarin@gmail.com

While grazing cattle, he watched the night sky and motion of celestial objects. While watching movement of planets he took an interest to create models that would simulate this movement. He designed and developed several models. The astronomical observation and design experience made it possible for him to move to London in 1743 to study astronomy. He began popular astronomy lectures for "Ladies and Gentlemen", with demonstration of his models and laboratory experiments. These lectures were so popular, that by the end of the 1740s Ferguson began to lecture not only in London, but also in Liverpool, Manchester and other cities. By the time, he became a well-known creator of the Planetariums, globes and other devices. In 1761, George III awarded him a lifetime pension and Royal Society elected him as a member 2 years later. Autodidact Ferguson was elected as the member of American Philosophical Society in 1770.

James Ferguson died on 16th November 1776 in London. He was the author of many articles in "Philosophical Digest" and "Magazine for Gentlemen", published several books. Two of them can be considered as principal: "Astronomy explained upon Sir Isaac Newton's Principles" (1756) and "Lectures on Select Subjects" (1760). These books were very popular due to absence of mathematics and simple Ferguson's style.

2 What Does the Mechanical "Ferguson's Paradox" Consist of?

It should be noted, authors of this article could not find an exact definition of the paradox. Ferguson formulated this principle by himself in a letter to his friend Cooper [3]. One evening in the club, a watchmaker told Ferguson that the dogma about "The Holy Trinity" is paradoxical, as it was impossible to consider the Trinity as a whole. Ferguson suggested that the watchmaker should solve another problem. "Suppose we will mesh one wide gear and three narrow ones with various numbers of teeth. Then we shall rotate the wide gear. How will the narrow gears rotate?" The watchmaker said that three narrow gears would rotate in the direction opposite from rotation of wide gear. Ferguson did not agree with his opinion. Next week he brought a model of planetary mechanism (see on the Fig. 1). The model consisted of a planet carrier and five gears: gear 1 was fixed on a revolute pair in the center and meshed with both gear 1 and three narrow gears 3, 4, 5. These three gears were also established on the planet carrier h on revolute pairs. Number of teeth on gears 1, 2 and 4 are the same. The gear 3 had fewer teeth, but gear 5 had more teeth than gear 1. While the planet carrier h rotated clockwise the gear 2 rotated in the same direction, the gear 4 moved, but did not rotate (therefore, it would move translationally along a circle). The gear 3 rotated in the opposite direction than the gear 2, but the gear 5 rotated in the same direction as the gear 2. The watchmaker rotated the model for a long time and accepted that he cannot understand and explain it. Then Ferguson said: "If we cannot explain a simple motion of the gears, how can we understand and explain "The Holy Trinity"?"

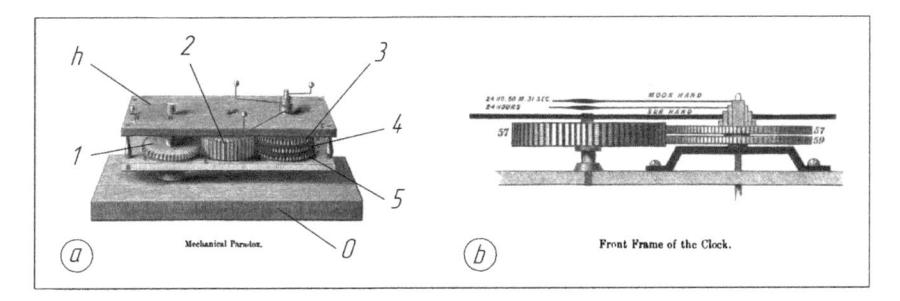

Analysis of Structure, Kinematic and 3D Modeling ... 185

Fig. 1 The mechanisms used by Ferguson in his planetariums [3]

3 Structural and Kinematic Analysis of Ferguson's Mechanism

Even today, "Ferguson's Paradox" amazes people. There is a scheme of the mechanism and its three rows of gearings on the Fig. 2. One problem which should be solved is fitting of the gears 2 and 3, 4 or 5 in the same center distance. At the same time, the gear 2 has one toothed ring with zero offset, and the gears have positive (4) and negative (5) offsets.

There is the graphic of kinematic analysis of the mechanism by Smirnov's velocity triangles method [4] on the Fig. 2, which clearly shows the direction of the mechanism links rotation.

Let us conduct structural studies of Ferguson's planar mechanism. Consider a differential version of the mechanism where a planet carrier and all of the gears can rotate about their axes. Number of the moving links of the mechanism $n = 6$. Number of one-mobility revolute pairs $p_1 = 6$. Number of two-mobility higher pairs $p_2 = 4$. Mobility of the mechanism calculated by the Chebyshev's formula [5] is 2:

$$W^{pl} = 3 \cdot n - 2 \cdot p_1 - 1 \cdot p_2 \\ = 3 \cdot 6 - 2 \cdot 6 - 1 \cdot 4 = 18 - 16 = 2. \qquad(1)$$

If we stop central gear 1 in the mechanism as it is shown on the Fig. 2, we will find a one-mobility mechanism where $n = 5$, $p_1 = 5$ and $p_2 = 4$:

$$W^{pl} = 3 \cdot n - 2 \cdot p_1 - 1 \cdot p_2 \\ = 3 \cdot 5 - 2 \cdot 5 - 1 \cdot 4 = 15 - 14 = 1. \qquad(2)$$

An equation which relates rotational movements of the links of the differential mechanism can be obtained from the Willis' formula [4]. We write this formula for all the gearing (when number of link $i = 3$, 4 or 5):

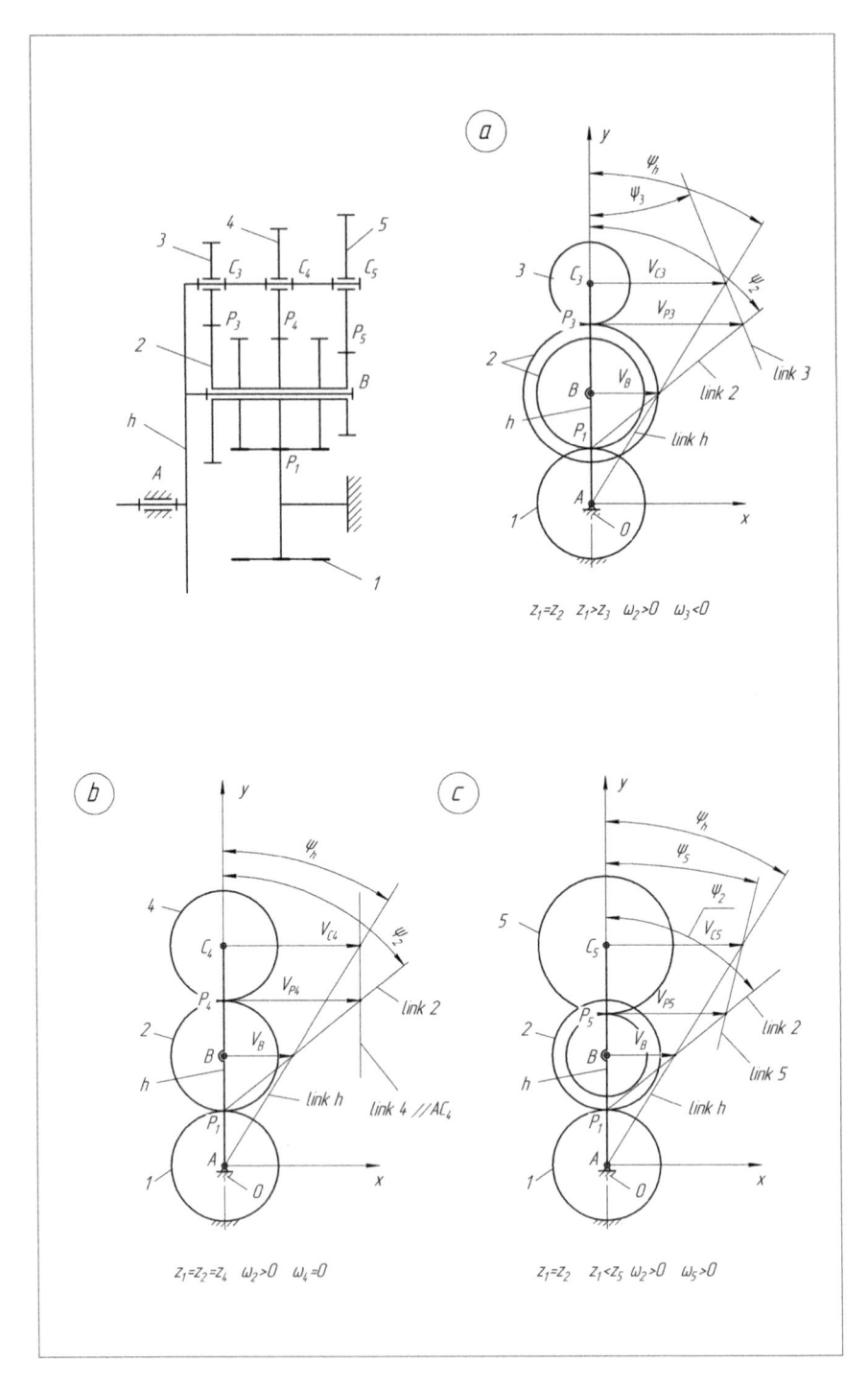

Fig. 2 Scheme of the Ferguson's mechanism and its kinematic analysis

$$\frac{\omega_1 - \omega_h}{\omega_2 - \omega_h} = -\frac{z_2}{z_1}; \qquad (3)$$

$$\frac{\omega_2 - \omega_h}{\omega_i - \omega_h} = -\frac{z_i}{z_2}. \qquad (4)$$

Then, for the gearing with gear i:

$$\frac{\omega_1 - \omega_h}{\omega_2 - \omega_h} \cdot \frac{\omega_2 - \omega_h}{\omega_i - \omega_h} = \frac{\omega_1 - \omega_h}{\omega_i - \omega_h} = -\frac{z_2}{z_1} \cdot -\left(\frac{z_i}{z_2}\right) = \frac{z_i}{z_1}; \qquad (5)$$

$$\omega_1 \cdot z_1 + (z_3 - z_1) \cdot \omega_h + \omega_i \cdot z_i = 0. \qquad (6)$$

For Ferguson's mechanism where the gear 1 is stopped and $\omega_1 = 0$, angular velocity of the gear i is

$$\omega_i = \left(\frac{z_1 - z_i}{z_i}\right) \cdot \omega_h; \qquad (7)$$

That is in this mechanism:

if $i = 4$, then $z_4 = z_1$ and $\omega_4 = 0$ and the link 4 makes circular translation (Fig. 2b);
if $i = 3$, then $z_3 > z_1$ the link 3 rotates in the opposite direction from the planet carrier h (Fig. 2c);
if $i = 5$, then $z_5 < z_1$, the link 5 rotates in the same direction as the planet carrier h (Fig. 2a).

4 Coaxial Ferguson's Mechanisms or Counting Mechanisms

Except the considered mechanism in the Ferguson's planetariums was used a coaxial mechanism like in the Fig. 1b. These mechanisms are called counting mechanisms in Redtenbacher's book [6]. There is a description of models from the Redtenbacher's collection, which were designed and manufactured in Karlsruhe Institute of Technology workshops. These models were prototypes for the workshop classes in the Statute of Moscow Craft School or SMCS (the name of BMSTU in those years) [7]. There are some models which can be called Ferguson mechanisms (Fig. 3) in the collection of Bauman Moscow State Technical University (BMSTU) department "Theory of Mechanisms and Machines" (TMM), two of which (Fig. 3b, c) are Ershov's models.

These models are similar to the Redtenbacher's models, but they have many differences in the shape of parts and their sizes. It can be seen in the Redtenbacher's drawings of the models (Fig. 4b, c) and in the photo (Fig. 4a) [8].

Fig. 3 Photos of the Bauman Moscow State Technical University models collection

In the introduction to [6] it was noted that the models from the collection are divided into four classes: "1. Machine parts; 2. Typical engineering products; 3. Transferring and motion conversion mechanisms; 4. Models of mechanism machines complexes. The first-class models demonstrate correct structural forms and proportions of machine parts. The second-class models are the image of the result of the production processes. The third-class models exactly show the real effect of different transmission mechanisms. The fourth-class models represent complete machines and mechanisms, which are difficult to show by drawings. This book presents just the third-class models."

Let us consider the models in Fig. 3b, c more detailed. The structure and kinematics of these models are equivalent to "Ferguson's paradox" model. Equations 1–3 and 6 are fully applicable to these mechanisms. When coaxial mechanism moves the center of the gear 3 is fixed so if the numbers of gear teeth

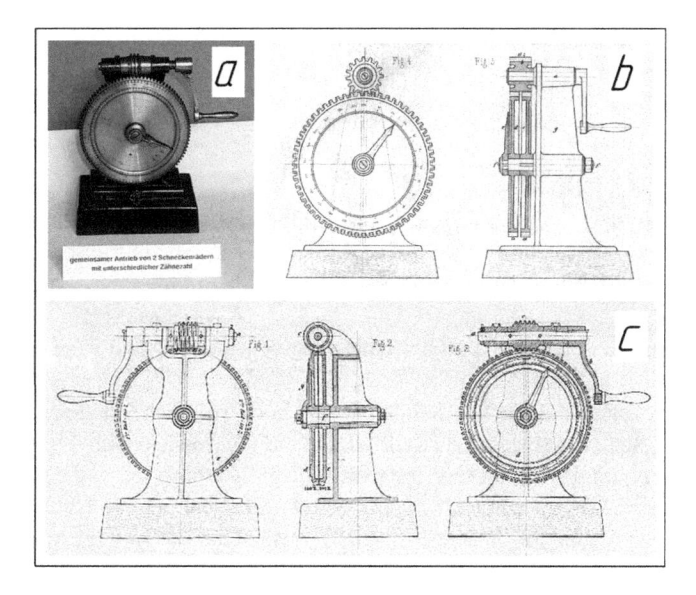

Fig. 4 Drawings and photo of the counting models from the Redtenbacher's book [6, 8]

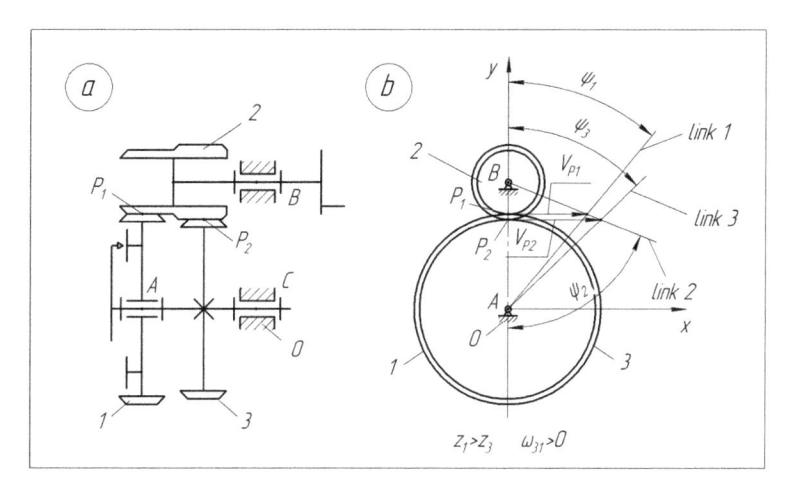

Fig. 5 The scheme of the coaxial disc counter with cylindrical gears

$z_1 = z_3$ the gear 3 is fixed. In the Fig. 5 there is a scheme of the mechanism 3b. "Disc counter: a is a rapidly rotating axis, rotations of which should be counted; b is a gear which is connected with a, the number of gear teeth $z_b = 15$; c and d are two gears; the first one has $z_c = 59$, the second one $z_d = z_c + 1 = 60$ teeth. f is an axis, which is mounted on the base g on a bearing. f is connected with gear c and pointer e, which indicates a scale on a gear d; d rotates on the axis f freely. The number of revolutions, which the axis a makes when the pointer has come full circle relatively to the gear d, is:

$$\frac{z_c \cdot z_d}{z_b} = \frac{59 \cdot 60}{15} = 236.$$

Therefore, the scale should be split in 236 parts so that one scale division corresponds to one rotation of the axis a'' [6]. Designations in the scheme (Fig. 5) and in the text [6] correspond to: link 1—d; link 2—b, a; link 3—c, g, f and e; stand 0—g.

The Fig. 6 is a diagram of counter with a screw gear (or with a worm gear), there are photos of it in Figs. 3c and 4a.

Redtenbacher told about the counter in his book [6]: "The model in Figs. 1, 2 and 3 (Fig. 4b), where a is an axis which rapidly rotates with handle b. It is necessary to count these rotations; the worm is rigidly connected to the axis a; d and e are two narrow worm wheels, which mesh with helical gear c. The worm wheel d has 100 teeth, but the wheel e has 101. The wheel e and the pointer g are connected to axis f; the pointer g indicates the scale on the gear d, which rotates on the axis f freely. When wheel c makes one rotation, wheels d and e rotate by one tooth. If a makes n turns, the wheel d will rotate by $\frac{360}{z_d} \cdot n$ degrees, but the wheel e will rotate by $\frac{360}{z_e} \cdot n$ degrees. The pointer g will rotate relatively to d by $\left(\frac{360}{z_d} - \frac{360}{z_e}\right) \cdot n = \frac{360}{10100} \cdot n$ degrees. After 10100 rotations the pointer g makes full circle on d. There is a scale with 10100 divisions. One division of this scale is one turn of the worm." Designations in the scheme (Fig. 6) and in the text [6] correspond to: link 1—d; link 2—b, a, c; link 3—g, f and e.

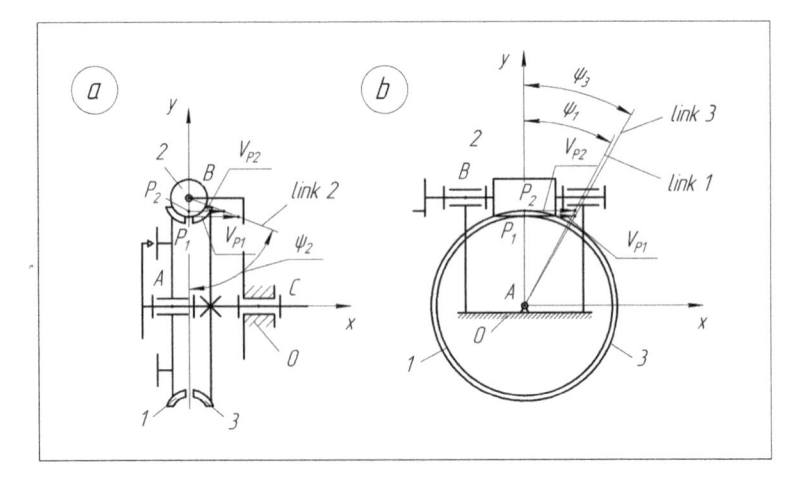

Fig. 6 The scheme of coaxial counter with "helical pinion"

5 Computer 3D Modeling of Ferguson's Coaxial Mechanisms

Today you can find photos and descriptions of the models [8, 9] in the Internet. There are physical and computer, virtual models [10, 11] of Ferguson's mechanisms. However, these models are designed to demonstrate "Ferguson's Paradox". In this case, a purpose of modeling is to obtain computer model of the real mechanism. It is a historical monument of science and technology. We want to create virtual object the same as in reality.

To create 3D-models of the mechanisms we used three-dimensional modeling system "КОМПАС-3D" ("COMPAS-3D"), which has integrated libraries with standard elements and tools for calculation and simulation of mechanical transmissions.

The original model of the counting mechanism was photographed, measured and disassembled. Modeling started with individual parts of the mechanism.

Elements in "КОМПАС-3D" were made from primitive objects such as box, cylinder, pyramid and so on. Individual elements of the detail were created by using the rotation operation or moving primitive objects along their rail (Fig. 7).

In this way, we developed model of axis, pointers, handles and bases. To draft models of gears we used library "Валы и механические передачи 2D" ("Shafts and mechanical transmission 2D"). By means of it, we calculated the geometry of transmission and obtained 2D model of the detail (Fig. 8a). Then the program generated 3D model (Fig. 8b). There is a model for assembling in Fig. 8c.

Nuts, washers and screws were selected from the library of standard elements. Their shapes and sizes were corrected.

We added missing details and set the limits, for example, coaxiality between axis, hole in the base and hole in the gear. Groove in the gear should coincide with the bore on the axis. There is a limit for gears rotation. They should have a definite transmission ratio. If the limits are correct, the parts will rotate as the original mechanism. There are assembled 3D models in Fig. 9.

Fig. 7 Development of the volume detail

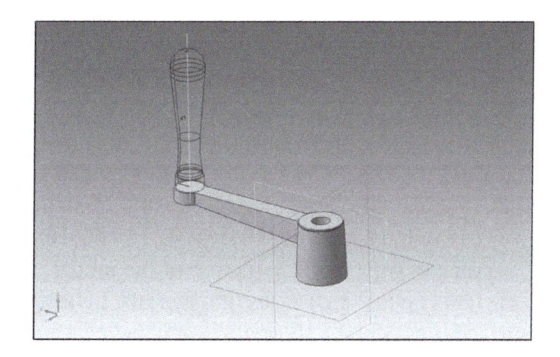

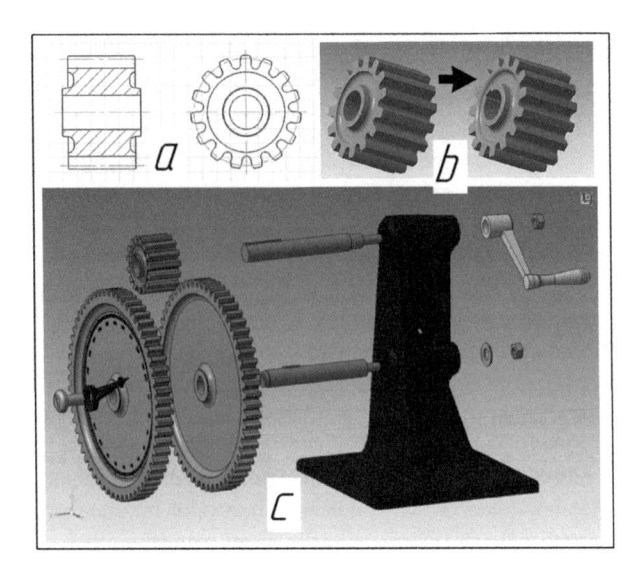

Fig. 8 Details of 3D model for assembling

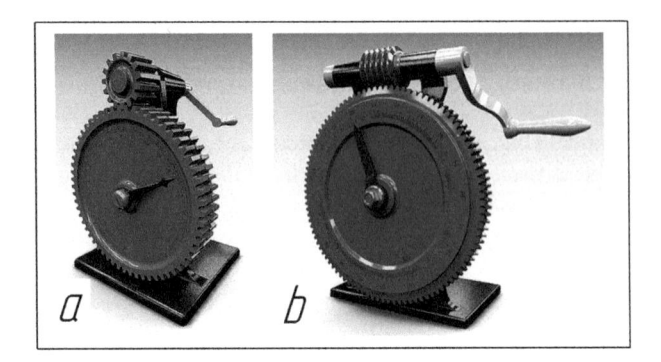

Fig. 9 Assembled 3D models of the mechanisms

Similarly, we created 3D model of the counting mechanism with worm. The worm and worm wheels were obtained from the "Валы и механические передачи 2D" module. Then we made the worm wheels thinner (Fig. 10).

Work of the mechanism can be demonstrated with animation library from "КОМПАС-3D". You can set rotation speed of the handle and watch the animation. You can also create more difficult animations with more details. For example, you can demonstrate the process of mechanism assembling.

You can get a realistic picture of the model with "Artisan Rendering" module. It considers optical properties of materials, lighting and environment. Every detail and surface is assigned a color or texture and processing method. The environment, type and intensity of lighting, background are chosen for the model as a whole. Then visualization process starts. There are results in Fig. 11a, b.

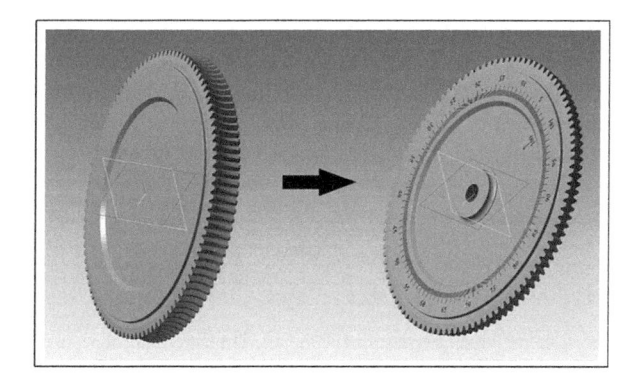

Fig. 10 Detail creating with the "Shafts and mechanical transmission 2D" library

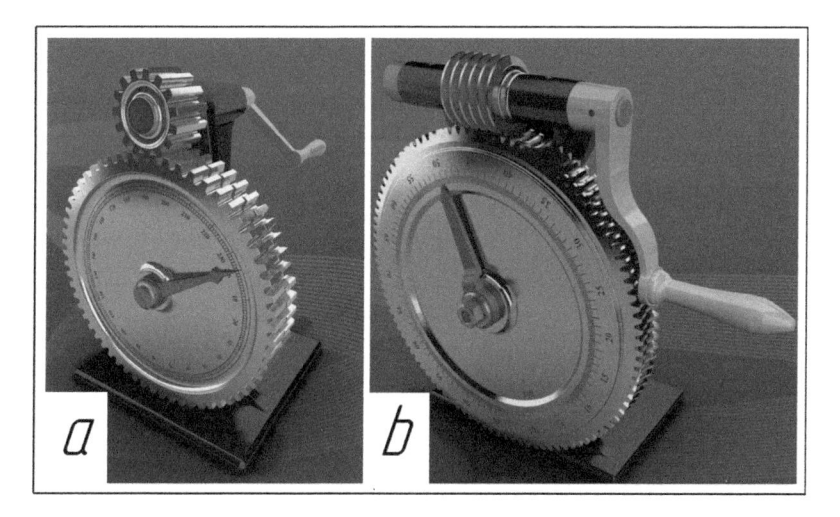

Fig. 11 3D visualizations

6 Conclusions

The paper contains brief biography of James Ferguson and descriptions of his "mechanical paradox". There are structural and kinematic researches of Ferguson's mechanisms. They describe two 3D models of counting mechanisms.

When Ferguson developed the models of celestial objects (Planetariums and Lunariums), he learned that the planetary mechanisms are necessary. He found a practical method of fitting of different gears in the same center distance. He applied the "angular correction" method, but this theory was developed only in the 20th century.

Discussed models of counters were recognized as monuments of science and technology in November 2011 (certificates #1040 and #1041). Authors developed their virtual models to popularize these monuments. Because they are small they can be placed on websites of virtual museums and other network resources. And people from all over the world will understand how they work.

References

1. Henderson, E.: Life of James Ferguson, F.R.S., LL.D (1867, 2nd edn. 1870)
2. Davenhall, C.: James Ferguson: A Commemoration. J. Astron. Hist. Heritage **13**(3), 179–186 (2010)
3. Henderson, E., Ferguson, J.: Life of James Ferguson, F.R.S. In a Brief Autobiographical Account, and Further Extended Memoir: Primary Source Edition, p. 532. Nabu Press (2013)
4. Tarabarin, V.B.: Theory of mechanisms and machines lections. 15th Lection. Planet. Mech. Kinematic (in Russian) (2015). http://tmm-umk.bmstu.ru/lectures/lect_15.htm
5. Tarabarin, V.B., Kuzenkov, V.V., Fursyak, F.I.: Laboratory Theory of Mechanisms and Machines Practice: Methodical Instructions for Laboratory Work on Discipline TMM, p. 96. BMSTU Publishing House (in Russian) (2009)
6. Redtenbacher, F.: Die Bewegungs-Mechanismen, p. 123. Bassermann, Mannheim (in German) (1857)
7. Golovin, A., Tarabarin, V.: Russian models from the mechanisms collection of Bauman university. Ceccarelli, M. (Ed.) Series History of Mechanisms and Machnes, vol. 5, p. 246. Springer (2008)
8. Redtenbacher Collection of Kinematic Mechanisms. University of Karlsruhe, Germany. Model: UK053 Worm and Two-wheel Drive Chain (2015). http://kmoddl.library.cornell.edu/model_metadata.php?m=347
9. Ferguson, J.: Mechanical Paradox Orrery. Amstrong Metalcrafts (2015). http://armstrongmetalcrafts.com/Products/ParadoxOrrery.aspx
10. Ferguson's paradox (2015). http://www.youtube.com/watch?v=UXUU0R-GSZ8
11. Ferguson mechanical paradox for clock motion work (2015). http://www.youtube.com/watch?v=vWNLDgU6Xl4

19th c. Olivier String Models at Cornell University: Ruled Surfaces in Gear Design

F.C. Moon and J.F. Abel

Abstract This paper describes a newly discovered collection of 19th c. Olivier string models for descriptive geometry at Cornell University, Ithaca, NY, USA. String models illustrating concepts of descriptive geometry were employed in the 19th and early 20th centuries as teaching and visualization aids for engineering, architecture and mathematics. The carefully crafted "Olivier models" are composed of finished wooden bases, brass armatures and colored threads and have an inherent beauty. They illustrate a variety of ruled surfaces—cylinders, hyperboloids, cones, conoids, planes and hyperbolic paraboloids (hypars)—and their intersections. The models not only served to enhance mathematical and drawing skills but also inspired designers and artists. Such inspiration is evident for both the design of mechanisms for machines and for the creation of 20th century sculpture. It is believed the models were copied after the historic original Olivier Models at Union College circa 1884.

Keywords String models · Ruled surfaces · Olivier · Gears · Hyperboloid · Hypoid · Cornell

1 Introduction

An impressive collection of 19th century geometric models for the teaching of descriptive geometry was recently discovered in the basement of Cornell University's Hollister Hall. These models use dozens of strings on metal supports to illustrate the intersection and tangency of solid objects. The models were re-discovered by former Chair of Civil and Environmental Engineering, John Abel [1, 2]. There are

F.C. Moon (✉)
Mechanical and Aerospace Engineering, Cornell University, Ithaca, USA
e-mail: FCM3@cornell.edu

J.F. Abel
Civil and Environmental Engineering, Cornell University, Ithaca, USA
e-mail: JFA5@cornell.edu

© Springer International Publishing Switzerland 2016
C. López-Cajún and M. Ceccarelli (eds.), *Explorations in the History of Machines and Mechanisms*, History of Mechanism and Machine Science 32, DOI 10.1007/978-3-319-31184-5_18

approximately 39 models each with a wooden box base and a brass medallion with the name "C.M. Clinton Model Maker Ithaca NY". Charles M Clinton [1834–1909] was a skilled machinist. He built sewing machines, engines and was the principal mechanic for the Ithaca Calendar Clock Co. In 1864 he began making high quality models, presumably for Cornell's College of Civil Engineering (CE) (Fig. 1).

In the 1913 CE catalog, there is a description of the Museums of the College of CE that mentions "over 50 brass and silk models made in the College after the Olivier models."

The Olivier string models are world famous. Theodore Olivier [1793–1853] was a student of Gaspar Monge [1746–1818] the founder of descriptive geometry upon which engineering drawing was based. Monge taught at the Ecole Polytechnique in Paris. Olivier was also one of the founders of Ecole Centrale in Paris and taught descriptive geometry at what is now the Musee des Art et Metiers (CNAM). Olivier had his models built in Paris by Fabre de Lagrange for CNAM [16]. Some universities in the US purchased their models from this Paris firm. A list of his models

Fig. 1 *Left* Hyperboloid with rotating *top* and *bottom* generator *circles*. *Right* Ruled surface cylinder and three hyperboloids; the inner called "kissing cones" [Cornell University Collection of Olivier String Models for Descriptive Geometry; Model CUCE 09;1884], [2]

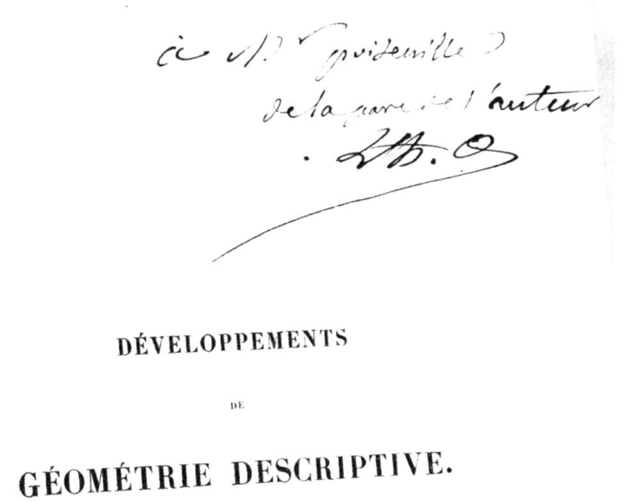

Fig. 2 Theodore Olivier Treatise cover with Olivier signature and dedication at the *upper right corner*; Cornell University Rare Books and Manuscripts Library

contains string models as well as gear models (See Herve [9]). Approximately 45 of his models are of the string type related to descriptive geometry.

Cornell University has a copy of Olivier's textbook on descriptive geometry in French and signed on the first page (Fig. 2). However the drawings accompanying the book are of an academic, mathematical nature and give no resemblance or clue to the construction of the string models he is so famous for.

How did Clinton copy the extremely intricate models of Olivier for Cornell and who paid for them? Records of the Trustees of Cornell 1884 show a request from Dean Fuertes of Civil Engineering for $780 to copy the Olivier models.

"In the Executive Committee meeting for Feb. 26, 1884, Prof. Fuertes appeared before the Board: "Asking for an appropriation of $780 to be expended in the reproduction of the Olivier Models in Descriptive Geometry." On Feb. 27, it was referred to "Committee on next year's estimates."

One clue is that Union College (older than Cornell) had a direct Paris copy of Olivier's set of approximately 41 models through the effort of a Prof William M. Gillespie circa 1855. Cornell President A.D. White mentions Union's President Eliphalet Nutt in his biography in connection with other matters. Could it be that Union College allowed Clinton to copy their models for Cornell? Another possibility is that Cornell professor John Morris of mechanical engineering, who was a graduate of Union College before coming to Cornell and was a student of Union's Professor William Gillespie, had brokered a deal with Union to copy their models.

String models illustrate a variety of ruled surfaces—cylinders, hyperboloids, cones, conoids, planes and hyperbolic paraboloids (hypars)—and their intersections; and through moveable frames that were Olivier's innovation, also demonstrate how these surfaces and intersections change with modification of the edge conditions. Olivier designed a set of prototype models around 1830 and had them built in Paris by the firm Pixii Pere et Fils. The carefully crafted "Olivier models" are composed of finished wooden bases, brass armatures and colored threads, the last kept in tension by weights suspended in the wooden base boxes—and in later copies by springs.

The Cornell models are perhaps direct copies of the prototypical personal models made to order by Olivier and acquired for Union College (Schenectady, New York) directly from the Olivier's widow in 1855 by Union Professor of Civil Engineering William M. Gillespie [1816–1868] [23]. Stimulated by these finds, the writers are undertaking not only a restoration of the Cornell collection but also further research on Olivier models and their influence on engineers, architects, artists and mathematicians. As of January 2016, twenty models have been restored. The photos of the Olivier models in this paper are all drawn from the Cornell collection, and some examples are shown in Figs. 1 and 3.

The Authors visited string model collections at Harvard, United States Military Academy (USMA) and Union College in order to compare their models with those recently discovered at Cornell. There are also model lists from the Paris Musee des

Fig. 3 Hyperboloid (*red*) and a hypar (*yellow*); *top* and *bottom* rings can be rotated [Cornell University Collection of Olivier String Models for Descriptive Geometry; Model CUCE 04; 1884]

Art et Metier, and the Merrifield Catalog of the Science Museum in London. Professor Fred Rickey of USMA has attempted a concordance chart of all these models. It is the Authors' belief that the model set closest to the Cornell-Clinton string models are the Union College Models representing Olivier's original collection.

1.1 Other Collections of Ruled Surface Models

Olivier arranged for copies of his prototype models to be built for sale to various universities in Europe and North America for use as teaching and visualization aids. Some copies were also directly acquired by European science museums [25]. Such manufacture and sale continued after Olivier's death, although the labeling on the models built in Paris still identified him as the inventor.

French-made copies of the Olivier string models were ordered by several universities in the United States, including Harvard, United States Military Academy and Columbia University. At Harvard University, an original collection of "above fifty" Olivier models were acquired in 1856 for the mathematics department, but by 1961 only four survived in attic of the school of engineering for transfer to the Collection of Historical Scientific Instruments [26]. Although depicted and well documented on the Waywiser website, all four of the physical models are currently in storage rather than on public display.

Others also built copies of the models; for example, Stone [23] reports that copies of Columbia University's Olivier models were made in the US for Princeton University.

In the UK, French copies were acquired by the London Kensington Museum now the Science Museum.

As teaching methods changed in the early 20th century, use of the Olivier string models diminished to the point of vanishing. Models that survived disposal became collections that are either displayed or in storage. The Cornell Collection has approximately 20 models on public display in Hollister Hall.

These elegant models have inspired artists as well as engineers. The British sculptor Henry Moore (1898–1986) mentioned seeing them in the Kensington Museum (now known as the Science Museum), and in the late 1930s he used strings as part of his works (The Royal Society [24]; the downloadable PDF file of [24] contains color photos of the Science Museum's Olivier string models and of Moore's string sculptures inspired by them). Among other sculptors who were inspired by surface and string models were Naum Gabo and Barbara Hepworth [25].

There are a number of mathematical model collections with solid representations of ruled surfaces including those at the University of Illinois, University of Arizona, MIT and at Cornell. In Germany the collection of models at the Mathematical Institute of Göttingen University is famous because some were made in plaster by

Felix Klein. Rigid models however do not have the deformability of the Olivier string models.

The Olivier string models appeared in the 1927 descriptive geometry book of Charles H Schumann of Columbia University as a text for engineering students [21]. It appears however, that the motivation was for the study of the intersection of surfaces in architecture and structural design and not the design of machines.

Although the Olivier models provide motivation for the use of the hyperbolic paraboloid surface for hypar structures, there is no evidence of the use of the hypar roof until the middle of the 20th century. In contrast, the use of hyperboloid surfaces in mechanical engineering have a century long tradition.

2 Hyperboloids and Ruled Surfaces in Gear Design

The study of ruled surfaces in machine design can be traced to the beginning of the 19th century. A survey of mechanisms in Lanz and Betancourt [12] Theatre of Machines book of 1808 does not show rolling hyperboloids as a means to transfer rotary motion from about one axis to another nonintersecting, nonparallel axis. Nor does a later French mechanisms handbook by J.-A. Borgnis [1818] show hyperboloid pairs [4].

One of the early treatises on the kinematics of machines is the 1841 work of Cambridge professor Robert Willis [1800–1875] [27]. (See Moon [15] also Notes Rec. Roy Soc London **57**(2), 209–230). Willis' book has a section on rolling cones and bevel gears in which he discusses hyperboloidal gears. (See p. 52, Section 67 as well as Section 47).

The later work of William Rankine (1887) mentions the work of Willis and has a number of citations to hyperboloids [17]. The French kinematics text *Traite de cinematique,* by Charles de Laboulaye (1861) explicitly cites the work of Theodore Olivier as well as mentions hyperboloidal gears (*Engrenages hyperboloidaux, p 286* [11]).

An early citation of rolling hyperboloids is in Ferdinand Redtenbacher's 1852 *Resultaten fur den Maschinenbau,* Tafel II, Fig. 9. This figure shows two gears as part of two tangent hyperboloids (Fig. 4) [18].

Since Franz Reuleaux [1829–1905] was a student of Redtenbacher, we would expect that ruled surfaces and hyperboloids would appear in his books and models. His early editions of *Der Constructeur, 1861 and 1865,* mentions "konoidische and hyperboloidische Räder" in a section on Zahnräder [19] (Chapter XIV, page 244, in the 1865 edition) More details appear in Reuleaux's 1875 and 1876 books *The Kinematics of Machinery,* translated by A.B.W. Kennedy. In Chapter II, there is a Section 13 on "Twisting and Rolling of Ruled Surfaces" [20]. Figure 32 shows a spiral surface rolling on another spiral surface. Figure 33 shows a cone rolling on a spiral surface (See Fig. 6 of this paper). On page 82 he writes "many ruled surfaces can roll upon one another". On page 83 he shows a drawing of a "hyperboloidal spur wheel" or what is called today a hypoid gear pair. He also gives two other references

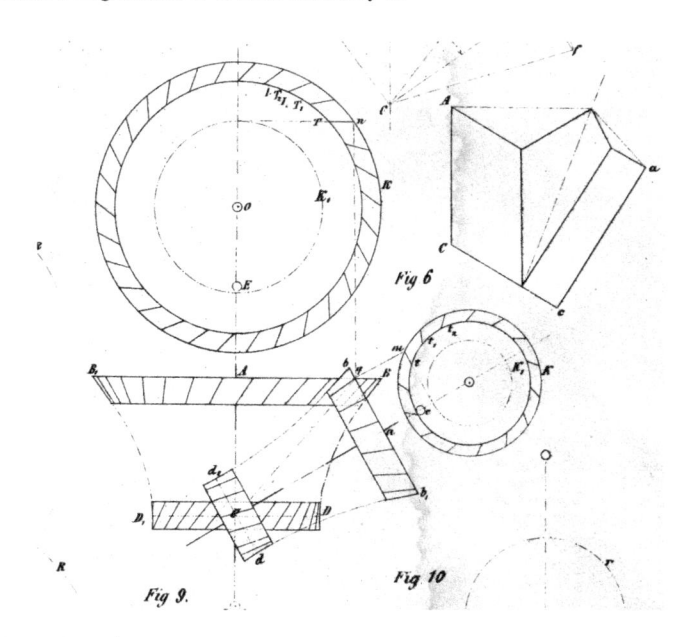

Fig. 4 Sketch from the work of Ferdinand Redtenbacher, *Resultaten für Maschinenbau*, 1852 showing crossed hyperboloid surfaces for gear design [18]

in the footnotes. Rolling bodies of ruled surfaces are discussed in Chapter XIII, Section 143; Fig. 365 shows a cone rolling on a spiral or helical surface. Also mentioned are a "pair of hyperboloids" and "hyperboloids of revolution".

Reuleaux made models of rolling ruled surfaces in his Berlin Collection. One shows rolling cones and another shows a pair of brass rolling hyperbolods. See (Fig. 5).

There is evidence that rolling hyperboloid pairs were common knowledge in the period before Reuleaux's theoretical work, as shown in the 1868 American mechanisms handbook of Henry T Brown, *507 Mechanical Movements*; Fig. 204 in Brown's book shows a diagram of such a pair with no extended explanation [5] (Fig. 6).

The influence of Robert Willis in England can be traced to Rankine who wrote about hyperboloids in mechanical design and to S. Dunkerly of University of Manchester whose 1904 text *Mechanism*, discusses rolling hyperboloids in Section 191 on "Skew-bevel Wheels" [6]. This text also shows a photograph of a string model of a hyperboloid in Fig. 53 in the 1910 Edition and a sketch of hypoid gears in Fig. 54. Another English kinematics professor was Alexander B.W. Kennedy who translated Reuleaux's kinematics theory book in 1876. Kennedy's own textbook of 1886, [10], described rolling hyperboloids in a section on "Skew-Gearing" (Section 68, p. 547).

In the American textbooks of MacCord [13] and early 20th century such as Cornell professors, Barr and Wood [3] and Albert and Rogers [22], the use of

Fig. 5 An example of one hyperboloid driving another by friction contact along opposing straight-line generators—a model from the Reuleaux Collection of Kinematic Mechanisms at the Deutsches Museum, Munich. DM-INV.NR 06/6231 [Photo F.C. Moon, 2001]

Fig. 6 Rolling of two ruled surfaces; a cone and a helicoid. From Kinematics of Machinery, Franz Reuleaux, 1876 [20]

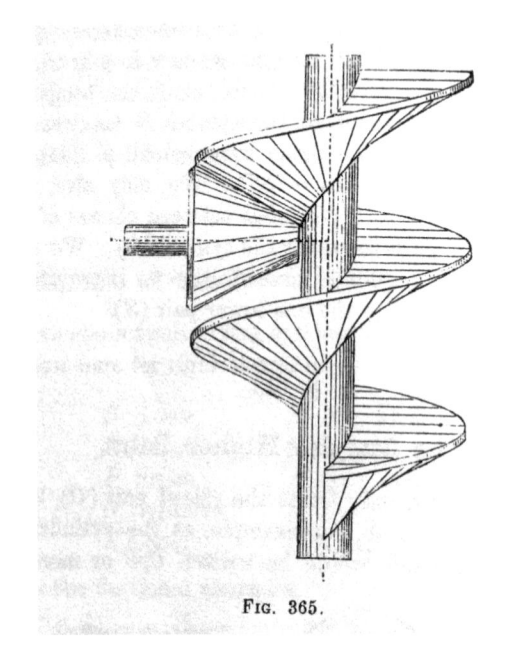

FIG. 365.

hyperboloid based gear pairs are mentioned and figures drawn. Charles MacCord of Stevens Institute in New Jersey also wrote a textbook in descriptive geometry in 1895 [14]. In this work he cited the influence of Olivier. He also has a chapter "On Warped Surfaces" (p 152, 1st edition) in which he described the construction of the conoid, hyperbolic paraboloid, and the hyperboloid of revolution. MacCord studied at Princeton University which was known to have had a set of Olivier models at the time.. MacCord in his kinematics book [13] acknowledged the influence of Willis, Rankine and Reuleaux.

Olivier himself bridged the fields of mathematics and mechanical engineering. He published two treatises on the geometry and working of gears (French engrenages), and the extended family of Olivier models included solid wooden models of gear mechanisms [9]. He was "the first who explored in depth the profiles of gear teeth for practical application" [9, p. 308], and many of these profiles were based on ruled surfaces represented by the string models.

Another design text that uses the hyperboloid is by Cornell University professors John Barr and Edgar Wood [3] *Kinematics of Machinery* (1916, 2nd ed) with an illustration in Fig. 101, page 97. In Canada, Professor RJ Durley of McGill University in Montreal published a textbook in 1903, *Kinematics of Machines* [7]. In Section 95, he described rolling hyperboloids and application to gears. McGill, like Cornell University had a very large collection of Reuleaux kinematic models. A number are sketched in Durley's book. The McGill collection was lost in a fire shortly after publication of Durley's book. Durley acknowledged the help of C.W. MacCord of Stevens Tech and J.H. Barr of Cornell (Fig. 7).

Fig. 7 Sketch of rolling hyperboloids from Cornell University textbook of Barr and Wood, 1899, 1911 [3]

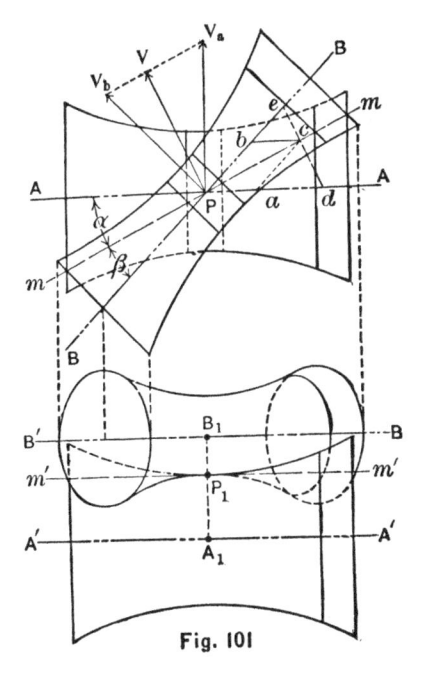

The connection between the use of ruled surfaces, gears and Theodore Olivier may be found in the design of machines literature. The work of Charles MacCord (Stevens Tech) who wrote a treatise *Kinematics* in 1885, [13], has a discussion of "Olivier involutes" (page 253) related to the shape of gear teeth. There are also a number of references to hyperboloids rolling on one another.

One of the most extensive treatises on the history of gears is the 1965 German work by H.-Chr. Graf v. Scherr-Thoss, *Die Entwicklung Der Zahnrad-Technik,* or *"The Development of Gear Technology"* [8]. He makes many references to Olivier's 1842 book on the *"Geometric Theory of Gears"*. Scherr-Thoss also discusses hyperboloid gears (p. 117, 128) and cites the US based Gleason Co which began making gears in 1865.

In the United States, The Gleason Co. of Rochester NY still manufactures skew bevel gears called *"Hypoid Gears"* in which the pitch surfaces are hyperboloids. As an aside, William Gleason, the founder of the company, had three children; the oldest of whom was Kate Gleason who was the first woman student to enroll in the Sibley College of Mechanical Engineering, at Cornell and the first woman member of ASME. Two other Gleason children also attended Cornell at the turn of the century. A son, James Gleason, became the President of Gleason Works. Gleason developed manufacturing techniques for hypoid gears in 1927 and has been a world leader ever since. Cornell had a set of Olivier hyperboloid models since 1884 in the College of Civil Engineering. We don't know if James Gleason ever saw or used these models. Nor do we know if he was influenced at all by the books of Cornell professors Barr and Wood [3] or Albert and Rogers [22].

With many examples of rolling hyperboloids in the 19th century kinematics literature from England, France, Germany and North America, it is a mystery as to why Olivier did not make a model of two tangent hyperboloids, given his own interest in the geometry of gears. Perhaps in making the string models he only thought of the mathematical questions of descriptive geometry. In the Olivier Paris wooden models of mating gears there are examples of pairs that appear to have skewed axes and hence would be part of hyperboloids (See e.g. Herve [9]).

Some of the Reuleaux kinematic models were purchased by Japan in the early 20th century and later copied by Japanese companies. One model that appeared in Japanese occupied Taiwan (shown in Fig. 8), shows a wire model of rolling hyperboloids with a hypoid gear pair. This model in on display in Tainan at National Cheng Kung University Museum [28].

Fig. 8 Wire model of rolling hyperboloids and hypoid gears in the Taiwan Collection of Kinematic Models [Courtesy Prof H.S. Yan] [28]

3 Concluding Remarks

While most recent studies of the Olivier models and other surface models have been written by historians of mathematics, the renewed interest in these 19th century string models has strong implications for engineering heritage. In civil engineering the structural forms of hypar warped geometry has been popular in late 20th century practice. Moreover, there is a sheer aesthetic pleasure in both their craftsmanship and appearance, as already recognized by sculptors and artists.

In mechanical engineering the use of the hyperboloid warped surface began in the mid 19th century and entered the practicing lexicon through many machine design textbooks in France, England, Germany and North America. The rediscovered Olivier models at Cornell complements the well-known kinematic model collection of Reuleaux, who was influential in establishing the importance of rolling of ruled surfaces such as the hyperboloid.

Acknowledgments The authors gratefully acknowledge the support of the Cornell University School of Civil and Environmental Engineering for the recovery and ongoing restoration of the collection of string models. Tim Brock, a staff member of the School of CEE, is carrying out the restoration of the string models, more than twenty of which have been restored as of this writing. In addition, the authors wish to acknowledge the collaboration of Fred Rickey, professor emeritus of mathematics at the U.S. Military Academy, in ongoing research regarding the Olivier string models. The curatorial staff of the Mandeville Gallery at the Union College Schaffer Library has generously afforded us the opportunity to examine the Union collection of Olivier's original models, currently largely in storage at that institution. Other sources include the Harvard Museum of Scientific Instruments, S. Schecter Curator, Elaine Engst Cornell University Archivist and Peggy Kidwell, Smithsonian Institution.

References

1. Abel, J.F., Oliva, J.G. (eds.): Special issue for the centenary of the birth of Félix Candela. J. Int. Assoc. Shell Spatial Struct. **51**(1) (2010)
2. Abel, J.F., Moon, F.C.: 19th c. String models for descriptive geometry: Possible inspirations for structural forms. In: Proceedings of the IASS-SLTE 2014 Symposium "Shells Membranes and Spatial Structures, (eds) Rayolando, MLRF. Brasil and Ruy MO Pauletti, Brasilia, Brazil (2014)
3. Barr, J.H., Wood, E.H.: Kinematics of Machinery. Wiley, New York (1899, 1911, 1916)
4. Borgnis, J.A.: Trait Complet de Mechanique Applique aux Arts: Composition des Machines, Paris (1818)
5. Brown, H.T.: Five Hundred and Seven Mechanical Movements. Brown Coombs and Co., New York (1868)
6. Dunkerley, S.: Mechanism. Longmans, Green and Co., London (1910)
7. Durley, R.J.: Kinematics of Machines. Wiley, New York (1907)
8. Graf von Seherr-Thoss, C.: Die Entwicklung Der Zahnrad-Technik. Springer, Berlin (1965)
9. Hervé J.M., Théodore Olivier, Distinguished Figures in Mechanism and Machine Science, Ceccarelli M. (ed.). Springer, Dordrecht (2007), pp. 295–318
10. Kennedy, A.B.W.: The Mechanics of Machinery. MacMillan & Co., London (1886)
11. Laboulaye, C.: Traite de Cinematique ou Theorie des Mechanismes, Paris (1849, 1864)
12. Lanz, P.L., Betancourt, A.: Analytical Essay on the Construction of Machines, Paris, London (1808)
13. MacCord, C.W.: Kinematics. Wiley, New York (1883)
14. MacCord, C.W.: Descriptive Geometry. Wiley, New York (1895)
15. Moon, F.C.: Franz Reuleaux: contributions to 19th C. Kinematics and history of machines. Appl. Mech. Rev. **56**(2) (2003)
16. O'Conner, J.J., Robertson E.F.: Théodore Olivier, MacTutor History of Mathematics, 2002. http://www-history.mcs.st-andrews.ac.uk/Biographies/Olivier.html
17. Rankine, W.: A Manual of Machinery and Millwork, London (1887)
18. Redtenbacher, F.: Resultate für den Maschinenbau. Verlag von F. Bassermann, Mannheim, Germany (1861)
19. Reuleaux, F.: Der Constructeur. Braunschweig (1861, 1864)
20. Reuleaux, F.: Kinematics of Machinery. MacMillan & Co., London (1876)
21. Schumann, C.H.: Descriptive Geometry, A Treatise on the Graphics of Space for Scientific Professions (1927)
22. Albert, C.D., Rogers, F.S.: Kinematics of Machinery. Wiley, New York (1931,1938)
23. Stone, W.C.: The Olivier Models. Friends of the Union College Library, Schenectady (New York) (1969)
24. The Royal Society: Intersections: Henry Moore and stringed surfaces, Catalogue to accompany an exhibition at the Science Museum, London, The Royal Society Publication DES2533, March 2012: https://royalsociety.org/~/media/Royal_Society_Content/z_events/2012/Intersections2012-04-04.pdf
25. Vierling-Claassen, A.: Models of surfaces and abstract art in the early 20th century. In: Hart, G., Sarhangi, R. (eds.) Proceedings of Bridges 2010: Mathematics, Music, Art, Architecture, Culture, Tessellations Publishing, 2010, pp. 11–18. http://bridgesmathart.org/2010/cdrom/proceedings/46/index.html

26. Waywiser, http://dssmhi1.fas.harvard.edu/emuseumdev/code/eMuseum.asp?page=collections (search for "Olivier"), Collection of historical scientific instruments. Harvard University
27. Willis, R.: Principles of Mechanisms. London England (1841, 1870)
28. Yan, H.S.: Antique Mechanism Models in Taiwan. National Cheng-Kung University Museum, Tainan, Taiwan (2010)

Mechanism of Laoguanshan Pattern Looms from Late 2nd Century BCE, Chengdu, China

Feng Zhao, Yi Wang, Qun Luo, Bo Long, Baichun Zhang, Yingchong Xia and Tao Xie

Abstract The development of technology is tightly interlinked to the introduction of the weaving loom, and in particular with the complex pattern loom. Probably, the most important testimony of this link is provided by the Chinese character *ji* with its various meanings including intelligence, human excellence, crossbow-trigger and weaving loom. Furthermore, the technological principles behind the pattern loom have been the key-inspiration behind very important breakthrough technological inventions throughout human history as e.g. the French Jacquard loom, telegram, then computers. However, the archaeological evidence has been missing. A new exceptional Han dynasty burial in Laoguanshan, Chengdu, southwest China in 2013, has revealed four models of wooden pattern looms dating back to 2nd century BCE, resulting in the first evidence of the use of pattern loom in the world. Hence, these finds provide the earliest existent archaeological material which is techno-

F. Zhao · Q. Luo · B. Long
China National Silk Museum, 73-1 Yuhuangshan Road, Hangzhou 310002, People's Republic of China

Y. Wang · T. Xie
Chengdu Museum, 18 Shierqiao Road, Chengdu 610072, People's Republic of China

B. Zhang (✉)
Institute for the History of Natural Sciences, Chinese Academy of Sciences, 55 Zhongguancun East Road, Beijing 100190, People's Republic of China
e-mail: zhang-office@ihns.ac.cn; 25724123@qq.com

Y. Xia
Zhejiang University of Technology, 182 Zhijiang Road, Hangzhou 310024, People's Republic of China

F. Zhao · Y. Wang · Q. Luo · B. Long · B. Zhang · Y. Xia · T. Xie
Jinzhou Conservation Center for Cultural Heritage, 142 Jingzhong Road, Jingzhou 434100, People's Republic of China

F. Zhao · Q. Luo · B. Long
Key Scientific Research Base of Textile Conservation, State Administration of Cultural Heritage, 73-1 Yuhuangshan Road, Hangzhou 310002, People's Republic of China

F. Zhao
Donghua University, 1882 West Yan'an Road, Shanghai 200051, People's Republic of China

logical speaking the point of departure for our present highly technological world. We have reconstructed the finds in order to identify the exact technological principles behind these pattern looms. Our results reveal that there are two systems for the movement of warp and weft resulting in a highly complex multi-shaft patterning principle with two different power transmission systems.

Keywords Laoguanshan pattern looms · Mechanism · Chengdu · China

1 Introduction

The exceptional find in 2013 of wooden pattern loom models excavated from a Han dynasty burial in Laoguanshan, Chengdu, southwest China, dates back to the second half of 2nd century BCE, and is the first evidence of the use of pattern loom [1].

A pattern loom is a weaving device with a set of shafts with heddles (or harnesses) where the pattern program was installed and used for continuous pattern repeats. The Chinese word for loom, *chi* or *ji*, represents the outline of a loom, indeed the ancient Chinese concept of crossbow-trigger and machinery par excellence [2]. The loom plays a vital role in the history of Chinese textile technology, but also in the world history of science and technology. However, until 2014, our knowledge of early history of the pattern loom was only based on textual mentions and excavated patterned textiles, but the pattern loom models excavated from Laoguanshan, Chengdu, document for the first time the early technology history. The loom models are thus the missing link between the technology of silk pattern weaving and the Chinese textile terminology. Moreover, and unexpectedly, they testify to the very advanced weaving technology of the treadle loom in 2nd century BCE Asia, a technology which is only known in the west a millennium later.

Four pattern loom models with some devices for warping, rewinding, weft winding, and fifteen painted wooden figures, each with their name written on the breast, probably representing weavers and weaving related workers, were found from tomb no. 2 of Laoguanshan in Chengdu, Sichuan province. The tomb chamber, 7.2 m long, 4.5 m wide and 2.75 m high, was made of painted wood and consists of five chests, one large on the top and four small below. The top one contained the coffin with a female corpse c.50 years old, named Wan Dinu according to a jade seal found outside the coffin, suggesting that the tomb had been robbed just after burial. The small northern chest below contains four pattern loom models while the other chests contain numerous lacquer objects. Based on the tomb style and a Western Han bronze coin, it was dated between the Emperor Jingdi's reign (157-141 BCE) and Wudi's reign (141-88 BCE) of the Western Han dynasty. No c14 dating was made.

L. 190   L. 186

L. 191   L. 189

Fig. 1 Northern chest with four loom models during excavation

The four loom models were made mainly of wood and partly of bamboo with cinnabar-dyed silk threads on the beams. Loom 186 is the largest model, with dimensions 85 cm long, 26 cm wide, 50 cm high, while the others, L.189, L.190 and L.191 are comparably smaller (Fig. 1).

2 L.186: Pattern Loom with Shaft

All four models had fallen apart when moved out from the burial, however, all elements are preserved and the looms could be reconstructed. Based on our reconstruction, L.186 has a loom body, 55 cm long, 24 cm wide and 12 cm high and supported by four legs, on which stands a loom castle, 25 cm long, 23 cm wide and 18 cm high, made of four slotted poles and containing all shafts. Two grilles made of bamboo, 10 cm high, each with 19 compartments, stand on either sides of the castle (Figs. 2, 3, 4, 5 and 6).

The warp beam and cloth beam stretch the warp on the loom, the former in the back to hold the warp, and the latter in the front to hold the cloth. They are both supported by two pairs of beam bears, 6.7 cm high, with ratchet on the right side. To accommodate a longer warp on the loom and have less change when the shed is made, a back beam or whip roll is placed on an extra frame, 24 cm long, at the back of the loom (Fig. 7).

The foundation shafts create the warp's natural and counter sheds, enabling the shuttle to pass and weave a tabby structure. On the inner sides of two front poles of the castle, there are two pulleys with two foundation shafts, about 10 cm high and 9.5 cm wide, which are operated by two treadles on the petal beam.

The pattern shafts installed in the castle is the most complex system of this loom. There are only five pattern shafts left, each hang by a shaft bar, 15.8 cm long, among 19 compartments of grille on the castle, meaning that in this section, the loom model was only an imitation of real pattern loom. During the weaving process, one side of the axle treadle held by the front poles is stepped and the other side turns up to push two sliding frames in the slotted roles upwards. Then the hook

Fig. 2 L.186 during excavation

Fig. 3 L.186 after conservation

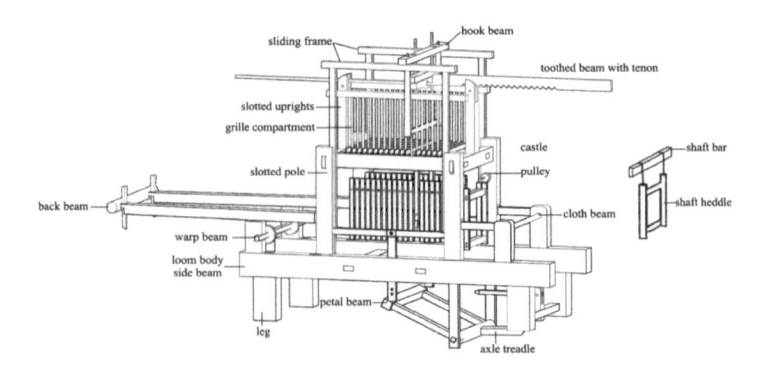

Fig. 4 Reconstruction of the structure of L.186 and terms

Fig. 5 Reconstruction of the structure of L.186

Fig. 6 Reconstruction of the L.186

beam on the top is also pushed up by two frames, to lift a shaft bar in a certain compartment thereafter a corresponding shaft. To make the selection of shafts, a toothed beam with two tenons can carry the hook beam moving in warp direction and stop at a certain position to select a pattern shaft. This movement could be operated by the weaver himself or his assistant.

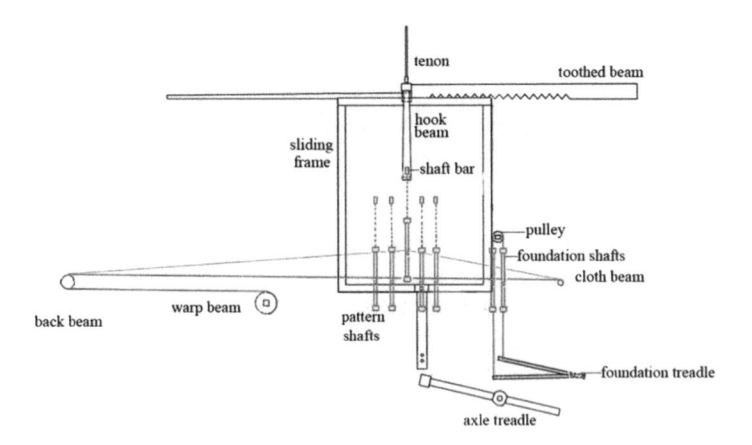

Fig. 7 Model of the movement of beams, shafts and treadles of L.186

3 L.189, L.190, L.191: Pattern Loom with Hook Rod

The other three loom models are smaller than L.186. Their power transmission method is also different from the L.186, with two hook rods on both sides instead of two sliding frames and a hook beam to lift the shaft bar. When the weaver steps on one side of the axle treadle, the other side pushes the hook rods upwards, and the hooks lift the shaft bar and the pattern shaft. So it is a linkage mechanism, the earliest evidence in China (Figs. 8, 9, 10, 11 and 12).

We therefore suggest a new terminology for these two types of four pattern loom models: hook-shaft pattern loom, the former type being a hook-shaft pattern loom with sliding frame, and the latter type being a hook-shaft pattern loom with hook rod.

Fig. 8 L.190 during excavation

Fig. 9 L.190 after conservation

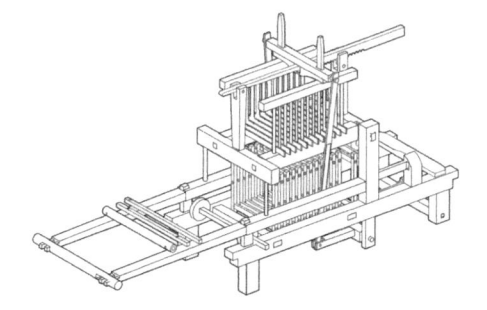

Fig. 10 Reconstruction of the structure of L.190 and terms

Fig. 11 Reconstruction of the structure of L.190

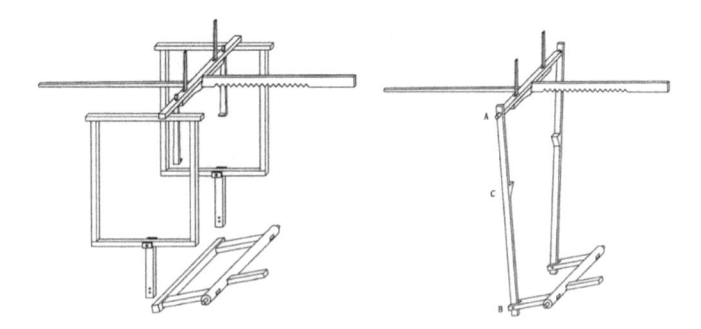

Fig. 12 Power transmission of L.186 (L) and L.190 (R)

4 The Movement of Treadles and Shafts

There are no textiles found on the looms, and no silk textiles found from the same period in the same area either. However, there are numerous polychrome woven silks, so-called *jin* silk [3], excavated from the Warring States, the previous period (5th–3rd centuries BCE), and the Early Han dynasty (2nd–1st centuries BCE) but in other areas. All these excavated *jin* silks used the same weave structure: warp-faced compound tabby, a particular and traditional weave structure for ancient China. The preservation of red silk thread with cinnabar and brown silk thread on these loom models with multi shafts strongly suggest that these looms were used to weave textiles similar to those with geometric pattern (Fig. 13) and bird pattern found at Mawangdui [4], Changsha in the Hunan province, and those with leopard pattern found in Fenghuangshan in the Jinzhou, Hubei province, both dated to the 2nd century BCE [5]. All the *jin* silks from this period including on this loom were probably made in a similar technique (warp silk yarn was dyed before being woven), and similar weave structure (warp-faced compound tabby) (Fig. 14). In this case of jin silk with red and brown colors, both red and brown warps in one group should enter one heddle of two foundation shafts alternately. Each group of warp should select either red or brown warp to enter one heddle of every pattern shaft according to the design. Thus, when one pattern shaft is lifted, each heddle should lift either red or brown warp to form the design. Here we use the jin silk with geometric pattern as an example for two foundational shafts and 24 pattern shafts to weave on the loom (Fig. 15).

Shed 1 for foundation: the foundation treadle 1 is stepped and foundational shaft 2 is lifted, the tabby shed is formed and the shuttle with foundation weft passes.

Shed 2 for pattern: the pattern treadle is stepped to push the sliding frame up. Pattern shaft 1 is lifted to make a pattern shed for the pattern weft, since the hook beam is moved with the toothed beam to the position of pattern shaft 1.

Shed 3 for foundation: the foundation treadle 2 is stepped and foundation shaft 1 is lifted.

Fig. 13 Geometric patterned *jin* silk from Mawangdui and drawing of the corresponding pattern

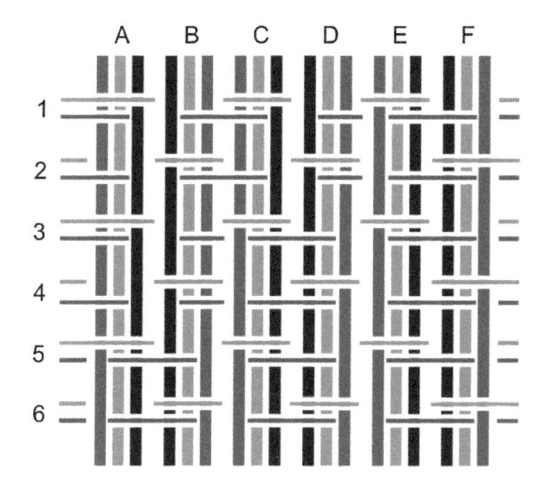

Fig. 14 Warp faced compound tabby = *jin*

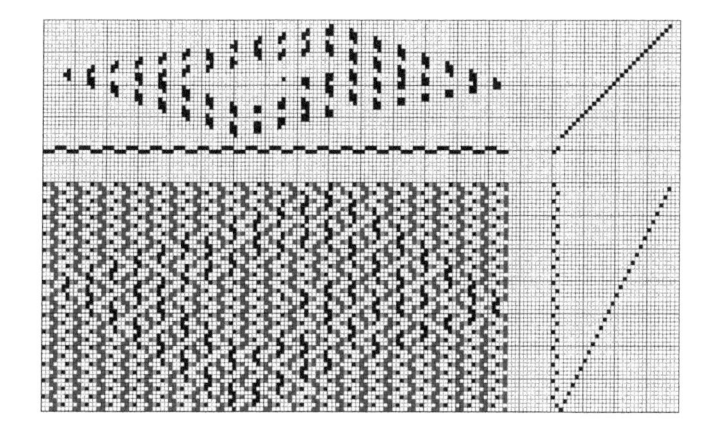

Fig. 15 Weaving plan for *jin* silk with geometric pattern

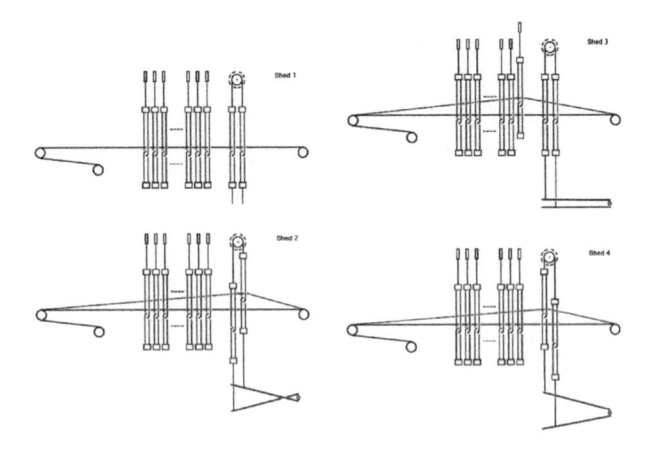

Fig. 16 Shedding process of this loom

Shed 4 for pattern: the hook beam is moved to the pattern shaft 2, then pattern treadle is stepped and pattern shaft 2 is lifted.

Then foundation shaft 1 again for shed 5, pattern shaft 3 for shed 6, foundation shaft 2 for shed 7 and pattern shaft 4 for shed 8, until pattern shaft 24 for weft 48, which is the end of one weave repeat. After the first repeat is finished, the pattern plan could be started again from 1–24 or 24–1, but in this case of geometric pattern, it should be started again from 1–24 (Fig. 16).

5 Conclusions

In archaeological excavations throughout the world, looms, loom models, elements of looms, and some loom illustrations have come to light, such as the horizontal two-bar loom model in a weaving workshop from tomb Meket-Re, Egypt, XIth dynasty, c. 2000 BCE, the vertical loom illustration from the tomb of Thot-nefer at Thebes, XVIIIth dynasty, c.1425BC [6], the warp-weighted loom illustrated in Greek vase painting of the 5th and 4th centuries BCE, and finds of loom weights from the Neolithic in Central Europe. Early loom types are also known in China, such as a back-strap loom found of the Liangzhu culture [7], and more in later periods [8]. All these loom types can be used to weave tabby or twill with 1-3 sheds, the weft is inserted manually, and they cannot produce programmed patterns. Another weaving technology with the capacity to include multiple weft systems and a mechanization of the loom through treadles came to light in China, but was so far only known as the oblique framed treadle loom for plain weave textile, the earliest example being a glazed pottery model from the Eastern Han dynasty, c.100-200 AD [9].

Several scholars, such as Flanagan [10], Crowfoot and Griffiths [11], Kuhn [12], and Riboud [13] have attempted to explain Chinese weaving technology based on ancient texts and on finds of archaeological patterned textiles. Most scholars such as Needham and James [14] attribute the invention of the pattern loom [15, 16] or draw loom [17, 18] to ancient China, but there is so far an on-going disagreement on how to define the technology of the type of Chinese-invented pattern loom, whether it was a multi-treadle and multi-shaft loom or a draw loom. The four newly discovered loom models excavated in Laoguanshan can now finally answer this question of technology history and give support to the multi-shaft loom type. Moreover, they reveal an innovation which scholars had so far never taken into consideration: the power transmission method to lift the pattern shaft.

The multi-shaft system is known solely from Chinese historical documents. In early Han dynasty, the 2nd century BCE, Chen Baoguang's wife wove patterned damask with 120 nie (metal sticks) which are probably pattern rods, suggesting the principle of pattern shafts [19]. Later during the Three Kingdom period, the 3rd century CE, a type of pattern loom for damask weave, with 50 treadles and 50 shafts or 60 treadles and 60 shafts, was in use in Fufeng [20], near present-day Xi'an. Both belong to the shaft pattern loom type.

Moreover, the axle treadle is also attested in later Han dynasty contexts. On an oblique treadle loom for plain weave from 1st to 2nd century CE, two treadles on a petal beam were connected to two arms at right angle on one axle, one pulling the axle to lift the shaft and release the warp tension, and the other pushing the axle to release the shaft but press the warp [21]. It is also a linkage mechanism, same as what used in this pattern loom.

Furthermore, several textile terms described in the late Han dynasty poem *Jifu fu* are similar to components of this pattern loom, such as *sheng* for warp beam, *fu* for cloth beam, *da kuang* for loom body, *guang* for shaft, *zhou* for treadle axle, *tu er* for cloth beam beam, *gao lou* for castle, *yu* for shuttle, *lu lu* for pulley [22].

The discovery of the pattern loom was made in Chengdu, with ancient name Shu, a city with an official *jin* silk workshop of the Qin and Han dynasties between the 2nd century BCE and the 2nd century CE, so this loom could be identified as a pattern loom to weave *jin* silk with warp faced compound tabby, in Shu area. Together with these pattern loom models, other weaving related tools, including warping devices, a warping board, silk re-winders and weft winders, and fifteen painted figures, including four male weavers and female weaving assistants, warping, weft winding, rewinding etc., were also excavated. This find is also providing relevant information about the proportions of the model and the real loom of a relationship of approximately 1:6 based on the size of the human figures, mostly 25 cm tall, and this was subsequently used for our reconstruction of the Han hook-shaft pattern loom.

The unique find of this 2nd century BCE hook-shaft pattern loom sheds new light on the early technology history and the transmission of innovations in the centuries BCE. It predates by several centuries the previous evidence of this weaving technology. It represents the missing technological link to the renowned Han dynasty *Shu jin* silks, which were found very often along the Silk Road, and traded across Eurasia. It demonstrates that the hook-shaft pattern loom was in use as early as the 2nd century BCE, and this suggests that it impacted on the inventions of the draw loom. It also emphasizes how Han dynasty China exported not only spectacular silk qualities and patterns to Central Asia and the Roman and late Roman west but also a different and more complex weaving technology. This archaeological find is thus significant to not only the Chinese history of silk and textile, but also the global history of science and technology. Undoubtedly, it is the earliest pattern loom model in the world, representing the most superb Chinese silk technique and reflecting the brilliant achievements in the human history. We anticipate that it was based on this shaft pattern loom that the draw loom was invented and introduced to the west, to Persia, India and Europe so the Chinese silk pattern loom made a great contribution to the world textile culture and technology.

Acknowledgments This project is supported by the Compass Plan, State Administration of Cultural Heritage, 2014. This a joint project by more than seven institutes, and we should thank the following people who helped and contributed in various ways: Professor Ziqiang Wang and Professor Yang Zhou of China National Silk Museum who studied the archaeological loom models; Mr. Mingbin Li and Mr. Yang Li of Chengdu Museum who assisted authors in dimensional measurement of the archaeological loom models; Ms. Jialiang Lu of Zhejiang Sci-tech university and Dr. Le Wang of Donghua University who drew the images of reconstituted *jin* silks; Dr. Hui Liu of Institute for the History of Natural Sciences, Chinese Academy of Sciences who provided historical documents. We also would like to thank Dr. Karin Frei of National Museum of Denmark for giving valuable suggestion and editing the manuscript.

References

1. Chengdu Institute of Archaeology: Jinzhou Conservation Center for Cultural Heritage. Archaeological report of Han tombs at Laoguanshan, Tianhuizhen, Chengdu, Sichuan. Archaeology **7**, 59–70 (2014)
2. Joseph, N.: Science and Civilisation in China. Cambridge University Press, Cambridge (1988)
3. Zhao, F.: Treasures in Silk: An Illustrated History of Chinese Textiles. ISAT/Costume Squad (1999)
4. The Archaeological Research Group of the Shanghai Textile Research Institute and the Shanghai Silk Industry Corporation. A Study of the Textile Fabrics unearthed from Han Tomb No.1 at Mawangtui in Changsha. Wenwu Publisher (1980)
5. Chen, Z.Y.: Han Tomb No.168 at Fenghuangshan, Jiangling. J. Archaeol. China **4**, 455–513 (1993)
6. Eric, B.: The Book of Looms: A History of the Handloom from Ancient Times to the Present. Brown University Press (1979)
7. Zhao, F.: Reconstruction of back-strap loom of Liangzhu. Southeast Culture **2**, 108–111 (1992)
8. Vollmer, J.E.: Archaeological and ethnological considerations of the foot-braced body tension loom. Studies in Textile History, Royal Ontario Museum, 343–354 (1977)
9. Zhao, F.: Reconstruction of axle-treadle loom in Han Dynasty. J. China Text. Univer. (English edition) **4**, 60–65 (1997)
10. Flanagan, J.F.: The origin of the drawloom used in the making of early Byzantine silks. The Burlington Magazine for Connoisseurs, 167–172 (1919)

11. Crowfoot G.M., Griffiths, J.: Coptic textiles in two-faced weave with pattern in reverse. J. Egypt. Archaeol. 40–47 (1939)
12. Kuhn, D.: Silk Weaving in Ancient China: From geometric figures to patterns of pictorial likeness. Chin. Sci. **12**, 77–114 (1995)
13. Riboud, K.: A detailed study of the figured silk with birds, rocks and tree from the Han dynasty. Bulletin de Liaison, CIETA **45**, 51–60 (1977)
14. James, J.M.: Silk, China and the drawloom. Archaeology **39**(5), 64–65 (1986)
15. Hu, Y.D. et al.: The development of Shu Silk Loom from a point of view of Dingqiao Loom: a field research report on the multi-treadle and multi-shaft Loom. Newsl. Hist. Text. Sci. Technol. China **1**, 50–62 (1980)
16. Tu, H.X.: Research and Reproduction of Silk Textiles from the Warring States Period. Donghua Univesity (1983)
17. Sun, Y.T.: The development of textile technology during the warring states to Qin and Han Dynasties. Res. Hist. **3**, 143–173 (1963)
18. Gao, H.Y., Zhang, P.G.: A Research of Development of Silk Weaving Machinery in Ancient China. China Textile Press (1997)
19. Xijing zaji (Miscellaneous records of the western capital) of the sixth century, attributed to Liu Xin or Ge Hong but probably by Wu Jun of Liang
20. Notes by Pei Songzhi for Wei Zhi (The Records of Wei Kingdom) of San Guo Zhi (The Records of Three Kingdoms)
21. Zhao, F.: Reproduction of Han Dynasty Oblique Treadle Loom. Cultural Relics **5**, 87–95 (1996)
22. Wang Yi. Jifu Fu (Rhapsody on women weavers) from the Eastern Han Dynasty

On the Warship by Ansaldo for Chinese Imperial Navy

Yibing Fang and Marco Ceccarelli

Abstract The paper presents a peculiar case of study of an attempt of technological transfer from Italy to China. It is remarkable the history of those conditions hat made China Imperial government to choose the Italian just established company Ansaldo for a first orders of a new warship. The history of this warship itself is peculiar and worthful to be discussed as an emblematic case of machine developments within international frames.

Keywords History of mechanical engineering · History of warships · History of Chinese imperial navy · History of Ansaldo

1 Introduction

It is well known that the relations between Italy and China in the scientific and technological exchanges can be traced back to the Venetian traveler Marco Polo (1254–1324), who came to China in 13th century, and the Jesuit missionary polymath Father Matteo Ricci (1552–1610), who produced considerable contributions in introducing the western scientific and technological knowledge into China in 16th century. However, the link between Italy and China in machinery industry during the period of Italian industrial revolution is seldom known.

It was in the second half of 19th century that Italy started a period of industrial revolution during which the modern Italian machine industry as well as its own modern machinery technology were established [1]. Some engineering enterprises

Y. Fang (✉)
Institute for the History of Natural Sciences,
Chinese Academy of Sciences, Beijing, China
e-mail: yibing@ihns.ac.cn

M. Ceccarelli
Laboratory of Robotics and Mechatronics, University of Cassino and South Italy,
Cassino, Italy
e-mail: ceccarelli@unicas.it

© Springer International Publishing Switzerland 2016
C. López-Cajún and M. Ceccarelli (eds.), *Explorations in the History of Machines and Mechanisms*, History of Mechanism and Machine Science 32, DOI 10.1007/978-3-319-31184-5_20

in the north Italy such as the Ansaldo Company developed rapidly into a engineering giant in the fields of shipbuilding and locomotive manufacture also thanks to the strategies for an economic nationalism supported by the government in the late of 19th century [1, 2]. At the same time, a process of the military development and technical modernization in China was promoted by some officials of the Chinese imperial government in the late 19th century. The establishment of the modern Chinese Navy as well as the import of European shipbuilding technology into China were important achievements during this process. Even though the connection between China and European countries in China's technical development in the late 19th century and the beginning of 20th century has attracted more and more attention in the field of historical studies of Chinese technology, most of previous studies are focused on the technology that was transferred from England or Germany, as pointed out in [3, 4]. Very limited attention has been paid on the link between Italy and China as probably due to the lack of the finding of historical archives or materials both in Italy and China.

In 2013 while carrying on a historical study of Ansaldo company which was considered a case study of the history of Italian machinery technology in 19th century, we found a booklet titled "the Specifications for the construction and Supply to the Chinese imperial Navy of a Naval destroyer" [5] (Fig.1), which was printed by Gio. Ansaldo Armstrong & Co in 1910. This pamphlet was printed as based on a warship-building contract between Imperial Chinese Government and Gio. Ansaldo Armstrong & Co in Genoa. It provides a special opportunity to discover unknown details on the historical connection between China and Italy on the machinery technology development in the beginning of 20th century.

This paper is focused on an explanation and interpretation of the information coming and developed by this contract for the warship of Ansaldo for the Chinese Imperial Navy (Fig. 1).

2 The Booklet and the Warship

The Ansaldo's booklet of 55 pages is a detailed specification for the destroyer which was constructed for Chinese Imperial Navy. Three parts are included in the content. The first part that is titled "Object" mainly gives a list of the principal data and characteristics of the ship and drive system, with the test of the drive system. The second part is a full description of the specifications for three parts of the ship, namely the complement part of the hull, the war armaments, and the drive system. The third part includes 6 attachments for more data and details of the equipment and auxiliaries machines that are annexed to the three principal ship parts.

The warship was a destroyer of 380 tons with 28.5 knots of velocity capability. The principal data of the ship are listed in Table 1.

Fig. 1 The booklet by Ansaldo on the warship for Chinese Imperial Navy [5]: **a** title page; **b** back cover

Table 1 The principal data of the destroyer built for the Chinese Imperial Navy [5]

Displacement normal	380 tons
Length	64.460 m
Breadth	6.096 m
Height	3.975 m
Medium immersion in full cargo	2.000 m
Machinery	VTE, 3 Thornycroft boilers
Power	6,000 Horse Powers
Max speed	28.5 kts
Fuel	Coal 50 tons, Oil 34 tons

3 The Backgrounds of Chinese Mission Visit to Europe

The beginning of Chinese modern Navy can be dated back to 1874 when the Imperial government decided to start a modern navy project [6]. During the first 10 years, 4 fleets were organized from South China to North China with 65 warships whose about forty-three were torpedo boats [7]. At the same time, a movement of self-strengthening was planned with the aim of establishing China's own technical ability in important industrial fields such as weapon manufacturing and shipbuilding as they were promoted by some officials of the imperial government. Around 20 modern engineering enterprises including a big shipbuilding company named Fuchou Shipyard was established by the end of 19th century [8]. When the Chinese Navy ministry was established in 1885, the most important navy force Peiyang fleet reached its golden time with 25 warships, most of which were imported from Germany and England, and only two were built by the Fuzhou Shipyard. When the Sino-Japanese war occurred in 1894, the Fuchou Shipyard produced 33 ships, most of which were for arming Chinese naval fleets, especially the Fujian and Guangdong fleet [9].

The defeat of China in the Sino-Japanese war of 1894–1895 marked a turning point for the development of the Chinese Navy as well as the China's shipbuilding industry. The Peiyang fleet suffered serious loss in this war with most of warships being sunk or captured. The Fuchou Shipyard also gradually lost the strong support from the government as a result of the disappointment that was caused by the defeat in the Sino-Japanese war. Finally the China's biggest shipyard was terminated in 1907 because of the serious financial difficulty [9].

It was not until 1909 that the Chinese Imperial government decided to reconstruct and expand the navy force when the younger brother of Guangxu Emperor, Zaixun (载洵) was appointed as the Navy Minister. A grand-scale plan of the reconstruction of navy was drawn up in July of 1909 with the financial support of 5,000,000 taels from the central government and other 11,000,000 taels from 18 Provinces government (Zhang 1982). Meanwhile a group of Naval Officials including the Prince Zaixun and Admiral Sha-Zhengbin (萨镇冰) were appointed as the Chinese Navy commissioners to be responsible for the implementation of the ambitious plan, especially for the purchase of new warships abroad. A investigation of the situation of the Chinese navy was made before the Navy commissioners left Beijing to Europe to investigate the navy and shipbuilding industry of European countries (Fig. 2) [6].

According to the statistical investigation at that moment, the warships of the Chinese Navy were only 32, and only six of them were made in China. It is noteworthy that there were 4 cruisers and 8 torpedo boats whose maximum loading of the biggest cruiser was 4300 tons and only 96 tons for the biggest torpedo boat. The manufacturers of the cruisers and torpedo boats were mainly from England, Germany and Japan. The other warships were mainly gunboats with most of which made by the Kawasaki Shipyard in Japan from 1905 to 1908 [6]. It is obvious that the situation of the battle ships of Chinese Navy in 1909 was unsatisfactory and even worse than that before 1894. It was under such situation that the Chinese Navy commissioners planned their visit to Europe in October of 1909.

The destroyer that was ordered as indicated in the Ansaldo's booklet was the first achievement made by the Chinese naval mission in Europe. It was on the 16th October, 1909 that the Chinese Naval mission, leaded by the Navy Minister Zaixun (载洵) and the Naval commander Sha zhenbing (萨镇冰) started its visiting tour to Europe. The Navy commissioners visited Italy, England, Austria and Germany from November of 1909 to January of 1910 [6] (Fig. 3).

Fig. 2 A Group photo of the Chinese Navy Commissioners before they embarked to Europe: the fourth one from *left* is Prince Zai Xun, and the third from *right* is Admiral Sha Zhenbing (The photo is from the collection at the National Library in Beijing)

Italy was the first destination of the Naval mission. Actually the Chinese navy commissioners visited Italy twice during their tour. They arrived at Genoa on about 19th of November, where Italy and Germany sent special representatives to welcome them [10]. After about 1 month in England, the Naval commissioners visited Italy for the second time from 24th to 30th of December, 1909. It was reported by in Reuter [10] that Chinese navy commissioners inspected the shipyard in Genoa on 25th of December and then they had an audience with King Victor Emanuel with whom they dined in Rome on 27th. Spezia and Venice were the last two destinations which were included in their schedule of the visits in Italy [11].

The photos in Figs. 2 and 3 at the National Library in Beijing show the historical relevance of the officer delegation with records of the technological tasks.

It is noteworthy that this was the first time for the Chinese Navy to send its officials to visit Italy. It is to note also that Italy was never listed in the warship suppliers for Chinese Navy before. Why did the Chinese navy commissioners choose Italy at that time? Two factors could be considered to be the reasons. Firstly, the remarkable development of Italy's shipbuilding industry in the late 19th century when Italy got into the process of industrial revolution [1], can be considered the primary positive condition, whose details are described in the next part of this article. Secondly, the active attitude of the Chinese ambassador in Italy played an important role on this matter. Before the Chinese Navy commissioners went to Europe, the Chinese Ambassador QianXun (钱恂) in Italy made an investigation of the situation with Italian navy as well as shipbuilding frames and he sent a report to the Emperor [12]. In this report, QianXun also mentioned that the Italian officials tried to persuade him to give attention in ordering a warship in Italy by describing

Fig. 3 Prince Zaixun inspected a Naval ship in Europe in December 1909 (The location of this photo is uncertain, but most likely that it was taken in Italy or England if we consider the date of the photo) (The photo is from a collection of National Library of China, Beijing)

three superiorities of Italian shipbuilding: the relative low cost because of the cheap labour; the relative short travel distance between the East and Mediterranean through the Suez Canal for saving the delivery time; the third benefit was that Italy had no territory in the mainland of Asia which may influence the relationship between China and Italy [12]. It is very likely that Qian Xun's report gave to the Emperor and his officials a clear knowledge of the situation of Italian shipbuilding and Navy, with indications that promoted to choose Italy as one of their cooperators in this Navy reconstruction plan.

It was Ansaldo Company and its shipyard that the Chinese Navy commissioners inspected during the 1909 mission. Soon afterward an order of a destroyer of 380 tons and 28.5 knots was issued to Ansaldo [5] which became the first shipbuilding order from China to Italy in the modern history.

4 The Activity of Ansaldo on Shipbuilding

The origins of the Ansaldo Company can be traced back to 1846 when the English engineer Philip Taylor and the Piemontese Fortunato Prandi established a workshop for the construction and repair of railroad machinery in Sampierdarena near Genova with a state loan [1, 13]. In 1853, the workshop was taken over by four new Italian entrepreneurs with a new name Giovanni Ansaldo & Co. It is to note that the four founders of Ansaldo company were all from Genoa with a vision to fulfil the need of a strong industry for power machines in the recently unified Italy.

Railroad machinery and shipbuilding were the two main areas of production in which the Ansaldo company was specialized in the 19th century [14]. It was in 1865 when the director of Ansaldo Luigi Orlando left the company to establish his own shipyard in Livorno that Ansaldo was reoriented to shipbuilding by the new director Carlo Bombrini. A shipyard was built beside the Ansaldo's machinery factory at Sampierdarena in Genoa. In 1876 the first ship that was fully built by Ansaldo, Royal Avviso Staffetta of 1,800 tons launched as shown in Fig. 4. However the development of the shipbuilding of Ansaldo in the first decade of its activities was not satisfactory because the company had to face a serious competition from the foreign manufacturers when a free-trade policy was adopted by the Italian government. In 1880s when the Italian state became the "agent of development" for Italy's heavy industries by adopting a protectionist policy, Ansaldo company got the state support to experience a rapid development, as pointed out in Row [14]. Thanks for the naval rearmament program sponsored by Admiral Benedetto Brin, the shipbuilding capacity of Ansaldo underwent a remarkable development since the late of 1880s. A new shipyard in Sestri Ponente near Genoa was bought in 1886. In 1889 Ansaldo successfully produced a steam engine of 19,500 horsepower for the Sicilia battleship, which became the most powerful engine in the world at that time. According to a news of a French Journal in 1896 [15], the cruiser Garibaldi which was made by Ansaldo for the Argentine Government got high appraisal for its low cost as well as very good operational performance and consequently the Spanish government ordered a large cruiser in Italy instead of in England [15].

By the turn of the century, two other Genoese ship-builders Attilio Odero and Giuseppe Orlando signed a cooperative agreement with Terni, the biggest steel producer in Italy. Thus an industrial combination—Terni-Odero-Orlando—was

Fig. 4 The first ship named as Royal Avviso Staffetta built by Ansaldo in 1867 [13]

formed which linked Italian most important steel and shipbuilding centres [14]. In order to avoid the dependence of the steel trust, Ansaldo joined with Armstrong's plant in Pozzuoli near Naples to form the joint-stock company Ansaldo-Armstrong & C. at the end of 1903, which became the largest metal-mechanical corporation in Italy [13]. The most important products of Ansaldo-Armstrong were the war ships, whose orders mainly came from the Italian Naval Ministry. Meanwhile foreign governments such as Argentina and Japan also became Ansaldo's clients [14].

It is interesting that the news of the merging event between Ansaldo and Armstrong was immediately reported by one of the most influential newspaper in Shanghai, China [16]. In this news, the shipyard of Ansaldo at Sestri Ponente was considered the most important shipbuilding yard in the Mediterranean as mentioned as "The amalgamation has for its object the development of the well-known shipyard of Ansaldo and Co. at Sestri Ponente, where men-of-war have already been built for the governments of Italy, Spain, Argentina, Brazil, Chile, and Romania. It was in this yard that the Rivadavia and the Moreno, recently bought by Japan from the Argentine Republic were built. The combination of English capital, and the world-wide experience and reputation of the Elswick firm, with the cheapness of Italian labour will render this the most important shipbuilding yard in the Mediterranean" [16].

Before the World War I, Ansaldo had grown into a first world class arms manufacturer. Ansaldo's shipyard was able to supply the Italian armed forces with warships and weapons which would have been imported 20 years before. On the other head, around one-third of the warships produced by Ansaldo was for foreign navies [17]. It was under such situation that the Chinese navy Commissioners chose Ansaldo to be one of the warship's suppliers.

5 Peculiarities of the History of the Warship Qing Po

It was a destroyer named Qing Po (鲸波) that the Chinese navy commissioners ordered to Ansaldo in Italy. The destroyer was a steel hull with 64.46 m long and 6.096 m wide. It had three Thornycroft boilers, and two propellers driven by a motor unit of 6,000 horsepower, that gave the capability to the vessel a speed of 28.5 nautical miles in full force. The destroyer was equipped with 2 cannons of 76 mm, 4 cannons of 47 mm, and three torpedo launchers. These data can be extracted from records in Ansaldo [5].

The Qing Po warship was the eleventh torpedo boat destroyer of the Soldato-class type which was built by Ansaldo (Fig. 5). Ten others were built for the Italian Royal Navy between 1905 and 1910, (Fig. 6) with an improved version of the Nembo-class destroyers built by the Pattison shipyard of Naples for the Italian Royal Navy between 1899 and 1905. Actually the initial design model of these two types of Italian destroyer can be traced back to the British destroyer of Havork style designed and built by the John I. Thornycroft & Company of Chiswick in England [18]. But the Italian shipbuilders improved the British design with more powerful

Fig. 5 The warship Destroyer Qing Po as under construction at the shipyard of Ansaldo [5]

armaments and nautical performance. In order to achieve this target, Italian engineers eliminated two small lifeboats on the both sides of the deck to obtain more rooms for equipping the armaments. Therefore four Torpedo tubes could be equipped on the boat while its British predecessor carried only two [19].

The warship Qing Po was ready on 6 December, 1912. Unfortunately the Chinese Qing dynasty was overthrown by the revolution of 1911. The new Chinese government was in serious financial difficulty at that time and could not pay the remaining loans of 56,000 Pounds [20]. Thus Italian navy took over the ship when Italy was involved into the World War I and gave it the new Italian name Ascaro [21]. Actually the Qing Po destroyer had a different fuel system from that of other members of Soldato-class with a fore boiler burnt oil and other two boilers were coal burners. The boilers of the ship was remained without any changes when the Italian Navy took over it. Meanwhile the armament of 2–3 in. and 4–47 mm guns was modified to the standard layout of Soldato-class with 4–3 in./40 Calibers guns [22]. During the WWI Ascaro was part of Destroyer Squadron IV. Downgraded to a torpedo boat in July 1921, Ascaro was disbarred in May 1930.

From the point of view of technology, the Italian Soldato-class destroyer can be considered one of the most advanced destroyers with relatively low cost in the early of 20th century. Therefore it was reasonable for Chinese imperial navy to order such kind of war ship in Italy. It was said that one other Soldato-class destroyer was

Fig. 6 One member of the Soldato-class destroyers Fuciliere at the port of Genoa, 1909 [5]

in construction at the shipyard of Ansaldo when the Chinese Navy commissioners arrived there. This visit gave to the Chinese officials very deep impression and made Prince Zaixun to decide immediately to issue an order for the same type of this destroyer to Ansaldo [19].

From the point of view of history, the Qing Po warship ordered by the Chinese navy not only represented the high level of the technological development of Italy's shipbuilding before WWI, but it also was a special turning point of the connection between Italy and China in the history of the engineering technology. Since then the Italian technology started to be considered in the Chinese programs. This created a new channel of technology transfer from western country to China and in the development of China's modern engineering industry in the early of 20th century.

6 Conclusions

This work reports the peculiar history of a warship that was ordered to the Italian company Ansaldo in 1909 by the China Imperial Navy. The motivations and conditions of such an order for which Italian shipbuilders was preferred are outlined by giving also an interpretation of the technological conditions and needs for a technological transfer to China frames. The history of the warship is also outlined to illustrate its characteristics and the peculiarity of this unknown event in the history of technological connections between Italy and China.

References

1. Fang, Y., Ceccarelli, M.: Peculiarities of evolution of machine technology and its industrialization in Italy during 19th century. Adv. Hist. Stud. **2015**(4), 338–355 (2015)
2. Fang, Y., Ceccarelli, M.: Findings on Italian Historical Developments of Machine Technology in 19th century Towards Industrial Revolution. In: The 11th IFToMM International Symposium on Science of Mechanisms and Machines, pp. 493–502. Spring International Publishing, Switzerland (2014)
3. Fang, Y.: The Hanyehping Company and the Iron and steel Technology Transplant in Modern China. Science Press, Beijing (2011) (in Chinese)
4. Wang, B.: Modern Railway Technology Transfer to China: the Case of Kiaotsi Railway 1898–1914. Shandong Jiaoyu Press, Jinan (2013) (in Chinese)
5. Ansaldo: Specifications: destroyer ship for Chinese State Navy. Gio. Ansaldo Armstrong & Co., Genova (1910) (in Italian)
6. Zhang, X. et al.: The Historical Data of Late Qing Navy. Haiyang Press, Beijing (1982) (in Chinese)
7. Zhang, G.: Yangwu Movement and Chinese Modern Enterprises, p. 24. Chinese Social Science Press, Beijing (1979) (in Chinese)
8. Rawlinson., John, L.: China's struggle for Naval development, pp. 1839–1895. Harvard University Press, Cambridge (1967)
9. Shen, C.: Fuzhou Shipyard, pp. 337–344. Sichuan Renmin Press, Chengdu (1987) (in Chinese)
10. Reuter (11-1909): The Chinese Navy Commissioners: A Great Reception. The North-China Herald and Supreme Court & Consular Gazette. November 27 1909
11. Reuter (12-1909): The Chinese Navy Commissioners. The North-China Herald and Supreme Court & Consular Gazette. December 31 1909
12. Qian, X.: Fifty Five Memorials to the Throne, pp. 138–142. Wenhai Press, Tai Bei (1970) (in Chinese)
13. Bagnasco, E., et al.: Storia dell'Ansaldo: 1. Editori Laterza, Roma (1994) (in Italian)
14. Row, T.: Economic Nationalism in Italy: The Ansaldo Company, 1882–1921. Ph.D. thesis, The Johns Hopkins University (1988)
15. Reuter. Latest Intelligence. The North-China Herald and Supreme Court & Consular Gazette. October 23 1896
16. The North-China Herald: Readings for the Period. In: The North-China Herald and Supreme Court & Consular Gazette, January 8 1904
17. Bagnasco, E., et al.: Storia dell'Ansaldo: 3. Editori Laterza, Roma (1996) (in Italian)
18. Gardiner, R., Gray, R.: Conway's all the World's Fighting Ships 1906–1921. Conway Maritime Press, London (1985)
19. Chen, Y.: The Warships in the End of the Qing Dynastic, p. 334. Shandong Huabao Press, Jinan (2012) (in Chinese)
20. Li, Z. (ed.): Historical Data of the Archives of Foreign Loans in the Republic of China, vol. 3, p. 525. Chinese Archive Press, Beijing (1989) (in Chinese)
21. Ma, Y.: New Annotations on the Historical Events of Modern Chinese Navy, p. 363. Lianjin Press, Tai Bei (2009)
22. Fracccaroli, A.: Itlian Warships of World War 1, p. 68. Ian Allan, London (1970)

Dynamic Analysis of an Ancient Tilt-Hammer

Umberto Meneghetti

Abstract In the first centuries of the second Millennium A. D., a sort of industrial revolution took place in Western Europe and many new machines were introduced. After this phase, there was stagnation in new machine appearance for a long time. Progresses in technology, however, did not cease but continued till the 18th century, when the Industrial Revolution exploded and the great growth in technology took place. Before this last event, the evolution of technology consisted more of updating of existing solutions than of introducing basic innovations. Nonetheless, from the early Middle Ages onwards engineers were able to advance many machines to a notable level of complexity and technological sophistication, showing considerable skill and power of observation. As an example, this paper describes a tilt hammer. The evolution of this machine is briefly depicted, and a particular hammer is then looked at, which presents non straightforward dynamical problems. A simple elastodynamic analysis is carried out and the operation of the machine is thoroughly explained.

Keywords Ancient hammer · Elastodynamic analysis · Tilt hammer · Industrial revolution

1 Introduction

Following the fall of the Western Roman Empire, technology suffered a general decline in Europe, even if ancient knowledge partially survived, because technology continued to be necessary for everyday life. Some centuries later, the renewal of economic activities stimulated a strong revival of technology to satisfying the increased population's needs. So, a sort of industrial revolution took place in Western Europe [3, 4, 9]. A number of machines were developed, some of them original or restored from Roman heritage, many imported from the East, these latter

U. Meneghetti (✉)
University of Bologna, Bologna, Italy
e-mail: umberto.meneghetti@unibo.it

sometimes improved by European craftsmen. Corn mills, ore crushers, forge hammers, fulling mills, and so on, were set up. Except for wind mills, these machines were usually driven by water wheels.

This phase of machines diffusion, was followed by a long period of stagnation, due to economic and political reasons. Progresses in technology, however, went on till the 18th century, when the Industrial Revolution exploded and the great growth in technology took place. Before this last event, the evolution of technology consisted mainly of updating of existing solutions rather than introducing of dramatic innovations.

In spite of this, machines' advancements in this period still showed the skill and power of observation from engineers of that time. As a matter of fact, they were able to advance many machines to a conspicuous level of complexity and technological sophistication. A noticeable example of this kind of progress is the tilting hammer considered in the present paper.

After a brief account of the hammer's evolution, a particular hammer is looked at, which presents non straightforward problems of dynamics. After a basic analysis of its operation, a simple elastodynamic model is proposed, which is then investigated by means of plain numerical methods. This analysis offers a better understanding of the way the hammer actually works and allows a critical analysis of its design.

2 Evolution of the Hammer

Primitive hammers were obviously operated by hand. To overcome force and energy limitations, they were soon mechanized, usually by a hydraulic wheel and a camshaft, as documented e.g. in [1].

Regarding the period we are interested in, for the first centuries of the Middle Ages no iconographical evidence is available, so we have to rely only on documents which testify the presence of many ironworks and mechanical hammers throughout Western Europe. However, very few technical data are available regarding power-hammers. In fact, descriptions of metallurgical processes do not include details on these devices. It seems as if they were considered irrelevant, or so well known that their description was superfluous. Figure 1 [5] shows two examples of hammers driven by hydraulic wheel and camshaft. Although these drawings were made later, they actually refer to the first centuries of the second millennium A. D. The drawings are schematic but they imply that the devices were rather primitive.

Since the mid XV century, on the contrary, there are accounts and pictures of hammers, from which we can infer that significant technological progresses had been made. Figures 2 and 3 [2], dating to the XVIII century, show in fact quite complex hammers accurately made by really skilled craftsmen, or rather, expert engineers.

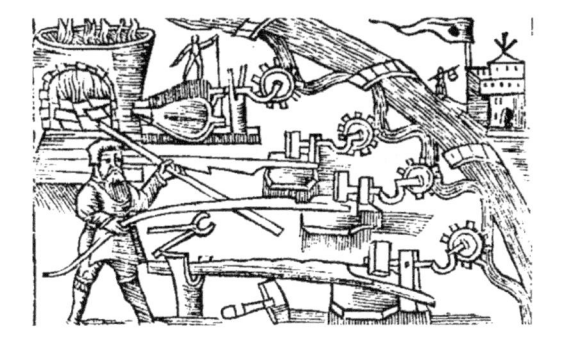

Fig. 1 Hammers driven by a hydraulic wheel

These hammers are of different kinds according to their specific application and the technology is evidently more advanced compared to previous examples. The hammer on the left in Fig. 2 is designed for the special case of forging anchors. It is quite different from the contemporary hammers shown on the right in Fig. 2, which are for copper manufacturing. The hammers in Fig. 3 on the left—for steel ingots—and Fig. 3 on the right—for tin-plates—are also different from each other and from the preceding ones. This differentiation in conformity with the application shows the highly advanced state of technology. In each case, the machines are rather complex and the execution looks very accurate, especially in comparison to the hammers in Fig. 1.

In some of the hammers shown in Figs. 2 and 3 the stroke is limited by an elastic restraint. In fact, the stem of the hammer is lifted by a cam and hits an overhanging elastic beam. The stem and beam bend and absorb elastic energy, which is then given back to the hammer. This way the stroke's duration is radically reduced and more hits per minute are obtained. The obvious drawback is the energy's waste due to hits and elastic deformation.

The same result was often achieved through a quite different solution, i.e. striking the hammer's tail against an underlying elastic restraint, as in Fig. 4.

The technique of the elastic stop proved to be the winning solution to obtain high energy and at the same time elevated stroke frequency, not achievable by gravity

Fig. 2 *Left* Hammer for forging anchors. *Right* Hammer for copper manufacturing [2]

Fig. 3 *Left* Hammer for steel ingots. *Right* Hammer for tin-plates [2]

Fig. 4 Tilt hammer and detail of camshaft and tail of the hammer [8]

force alone. In fact, this device was actually adopted up to the mid XIX century, and in some places even later.

As it is clear, in these hammers it is taken advantage of their elastodynamic behavior. It therefore seems interesting to carry out an elastodynamic analysis of such a mechanism by current engineering methods, which their designer were not able to do. A better understanding of the actual dynamic behavior of this machine could be worthy of note for a more in-depth knowledge of history of Theory of Machines and Mechanisms.

3 Dynamic Analysis of the Hammer

3.1 Working Process

We consider the hammer in Fig. 4 [7, 8]. Actually, the plant includes a hydraulic wheel that, through reducing gears, drives a flywheel and a camshaft with three sets of cams. Each set controls a different size hammer, the one in the figure is the biggest one.

In [8], with regard to this hammer it is described as follows: "When the cogs … strike the tail of the hammer suddenly down, … [this] strikes upon a support n, which acts to stop the ascent of the head of the hammer c, … but as the hammer is thrown up with a considerable velocity as well as force, the effort of the head to continue its motion, after the tail strikes the stop n, acts to bend the helve L ... and the elasticity of the helve recoils the hammer down upon the anvil with a redoubled force and velocity to that which it would acquire from the action of gravity alone ... The stop n ... bend[s] ... every time the tail of the hammer strikes upon it, and this aids the recoiling action very much."

The statements are quite correct, but they are exclusively descriptive and intuitional, not scientific. In particular, the assertion that the elasticity redoubles force and velocity is imprecise, because the whole energy is transmitted to the hammer during the contact with the cam and cannot be increased by a passive element like a spring or an elastic beam. Actually, basic concepts of Mechanics such as inertial force, kinetic energy, momentum, and so on are not made use of to explaining the working process. The feeling is that these concepts were unknown both to the readers and the writer, so we can reasonably infer that also the designer proceeded only intuitively on the basis of wholly empirical know-how.

3.2 Numerical Data and Initial Conditions

Referring to the scheme of Fig. 5, the main data of the hammer are: mean cam radius $R_c = 872$ mm; external cam radius $R_e = 915$ mm; distance between tail's end and center of motion $R_t = 1250$ mm; distance between center of motion and center of mass $r = 1146$ mm; distance between head of the hammer and center of motion $b = 1808$ mm; height of hammer's helve $h = 276$ mm; mass of the helve $m_h = 161$ kg; mass of the head of the hammer $m_b = 165$ kg; total mass of the hammer $m = 326$ kg; moment of inertia with respect to the center of motion

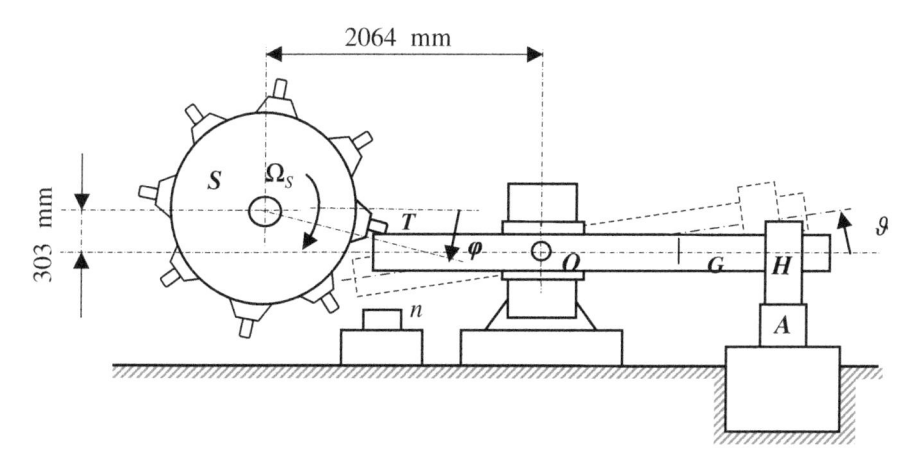

Fig. 5 Scheme of the hammer of Fig. 4. O is the center of motion, G the center of mass, H the head of the hammer, T the tail, A the anvil, n the stop, S the camshaft

$J = 730$ kg·m^2; angular speed of the camshaft $\Omega_S = 1.964$ rad/s; frequency $n = 150$ strokes per minute. Most of these data are reported in [8], but some dimensions were inferred from the scale drawings.

Referring to Fig. 5, it results that a cam hits the hammer's tail when angle ϑ is zero and the angle φ defining the cam position with respect to the horizontal has a corresponding value $\varphi_0 = 10.4°$. A pure geometrical-kinematic analysis shows that the cam leaves the hammer's tail at an instant t_1 from the very beginning of the rise, after a camshaft rotation $\varphi_1 = \Omega_S t_1$, when angle ϑ has a corresponding value ϑ_1. Values of φ_1 and ϑ_1 can be found through graphical-numerical analysis. We obtain $\vartheta_1 = 11.85°$, $\varphi_1 = 27.67°$, therefore it results $\varphi_1 - \varphi_0 = 17.27°$, $t_1 = 0.154$ s.

Without a stop, the maximum value of ϑ should be 19.1°, the resulting rise of the hammer 591.6 mm and the hammer's angular speed when striking the anvil $\dot{\vartheta}_3 = -1.82$ rad/s, linked to a linear velocity of the head $v_s = 3.29$ m/s. Therefore $\vartheta = 0$ after $t_2 = 0.603$ s, and the total duration of the cycle is $t_c = 0.757$ s, corresponding to a frequency of 79.3 strokes per minute (spm). Since the actual frequency is $f = 150$ spm, we can deduce that the stop was added to obtain higher frequency, as will be illustrated in the next sections.

3.3 Elastodynamic Analysis

Elastodynamic analysis is carried out taking into account the elasticity of the mechanism's members through a lumped parameters model. More specifically, we set two springs with stiffness k_c and k_s, see Fig. 6. Spring k_c models cam and helve's elasticity, while k_s models helve and stop's elasticity. Numerical values of k_c and k_s cannot be evaluated, because many dimensions, the kind of wood and contributions of local deformability are unknown, so we will try to assess these values by an indirect method, as will be shown later.

The hammer's cycle proceeds through three steps: cam-tail contact, tail-stop contact, free drop of the hammer. Geometrical investigation shows that the free motion from the end of cam-tail contact and tail-stop hitting is so small that it can be ignored.

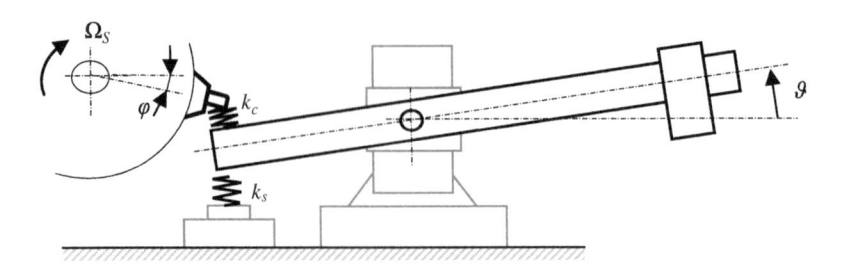

Fig. 6 Elastodynamic model

During the first step, the equation of motion—ignoring the shifting of the contact point between cam and helve—is:

$$J\ddot{\vartheta} + c_1\dot{\vartheta} + mgr\cos\vartheta + k_c[R_c(\Omega_s t - \varphi) - R_t(\vartheta - \vartheta_1)]R_t\cos\vartheta = 0, \quad (1)$$

where c_1 takes into account of passive resistances for this step.

As previously stated, the values of c_1 and k_c are unknown; however, numerical integration of Eq. (1)—e.g. using SIMULINK®—is very swift, so many different tentative values can be easily introduced. The values $c_1 = 1500$ N·m·s, $k_c = 65400$ N·m^{-1} provide consistent results, i.e. a rotation of the helve $\vartheta_1 = 11.85°$ in $t_1 = 0.16$ s, corresponding to a camshaft rotation $\varphi_1 - \varphi_0 = 18°$, just a little higher than the kinematic value of 17.27°, which is plausible; the final angular speed is $\dot{\vartheta}_1 = 2.42$ rad·s^{-1}.

The second step corresponds to the contact between tail and stop. The equation of motion is:

$$J\ddot{\vartheta} + c_2\dot{\vartheta} + mgr\cos\vartheta + k_s R_t(\sin\vartheta - \sin\vartheta_1) = 0. \quad (2)$$

where c_2 takes into account of passive resistances for this step.

Integrating Eq. (2), the initial conditions are $\vartheta = \vartheta_1$, $\dot{\vartheta} = \dot{\vartheta}_1$. We call the values of ϑ, $\dot{\vartheta}$ at time t_2 ϑ_2, $\dot{\vartheta}_2$ corresponding to the end of the phase.

In the last step, which is the hammer's free motion up to hitting the anvil, the equation of motion—ignoring frictional resistances—is simply

$$J\ddot{\vartheta} + mgr\cos\vartheta = 0, \quad (3)$$

with initial conditions $\vartheta = \vartheta_2$, $\dot{\vartheta} = \dot{\vartheta}_2$. The cycle ends at time t_3 when $\vartheta = 0$.

The hammer's head rise is $H = b\sin\vartheta$ and the total duration of the cycle is $T = t_1 + t_2 + t_3$. It is known [8] that the actual rise of the hammer's head is about 432 mm and period T is a little less than 0.4 s. Since the numerical integration of Eqs. (2) and (3) is very swift, it is easy to find by try the unknown values of c_2 and k_S corresponding to the actual values of H and T.

Attributing the following values to $k_S = 1.1 \cdot 10^6$ N·m^{-1} and $c_2 = 14000$ N·m·s/rad, the result is $t_2 = 0.072$ s, $t_3 = 0.13$ s; therefore, having previously found $t_1 = 0.16$, it follows that $T = 0.36$ s. The period corresponding to the actual frequency of 150 spm is 0.40 s, but it is to be expected that the helve reaches the start position on the anvil just a little before the cam hits it again, so the value $T = 0.36$ s is consistent. This also means that $\dot{\vartheta}_2 = -1.25$ rad/s.

The hammer's angular speed when striking the anvil is $\dot{\vartheta}_3 = -1.90$ rad/s, corresponding to a linear velocity of the head $v_s = 3.44$ m/s, a worthy value for the use for which the hammer is designed.

3.4 Discussion

The value $k_c = 65400$ N·m^{-1} for the cam-and-tail stiffness is very low, but not impossible. It could be due to the pliability of the connection between cam and camshaft.

The value $k_S = 1.1 \cdot 10^6$ N/m for the tail-and-stop stiffness is totally reliable. In fact, the stiffness of the helve from one end to the other, considered as a beam, is about $2.2 \cdot 10^6$ N·m^{-1}. The helve being in series with the stop, the reported value is reliable.

The maximum value of ϑ is $\Theta = 14.13°$ and the corresponding rise of the hammer is $1808 \times \sin\Theta = 441$ mm, reasonably close to the claimed value $H = 432$ mm.

To assessing the advantage of the use of the elastic stop, we can try to evaluate the behavior of the hammer working at the same frequency, that is 150 spm, but without the stop. We assume the same angular speed of the camshaft and the same number of cams, but a different configuration of cam and tail contact, so that the hammer's angular speed when leaving the cam is nearly zero and angle ϑ is approximately 9°. Under these conditions, the cycle duration is approximately 0.4 s, but the hammer's angular speed when hitting the anvil would be about 1.2 rad/s instead of 1.90 rad/s. Therefore, the energy of a strike is only 40 % of the energy using the stop.

4 Conclusions

An ancient mechanism was analyzed through modern but ordinary engineering means and its operational behavior was plainly illustrated. In particular, the advantage of using an elastic stop was ascertained and approximately quantified.

It is worth noting that the designers of this kind of mechanisms, as of all ancient mechanisms, did not use any theoretical means, even when these means were at their disposal. In fact, it has been already noted [6] that J. Smeaton (1724–1792), quoted as the celebrated designer of the hammer, was a member of the Royal Society of London. He was certainly acquainted with Newton's laws, but it sounds as if he did not consider using this knowledge for his engineering works [7, 8].

This general lag between scientific results and technological applications is quite amazing for us, but it was perhaps normal in the past, when technological knowledge was acquired "in the field" and transmitted directly only from master to apprentice.

In fact, before the XX century, European craftsmen and engineers had only a rough awareness of the physical and mathematical laws governing technology. But working by instinct, wisdom, trial and error, and perseverance, they were able to develop many machines to astonishing levels of complexity and technological sophistication—like the case of the tilt hammer—showing a remarkable skill and power of observation, transforming the world [3] and becoming the primary builders of the Industrial Revolution.

References

1. Biringuccio, V.: De la Pirotechnia (1540). Reprint, Dover, ISBN: 0486446433 (2005)
2. Diderot, D., Le Rond d'Alembert J.B.: Encyclopédie, ou Dictionnaire Raisonné des Sciences, des Arts et des Métiers (1751–1772)
3. Gies, F., Gies, J.: Cathedral, Forge, and Waterwheel. HarperCollins, ISBN: 0 06 092501 7 (1995)
4. Gimpel, J.: La Révolution industrielle du Moyen Âge. Seuil, ISBN: 2 02 054151 3 (1975)
5. Magnus, O.: Historia de Gentibvs Septentrionalibvs (1555). Reprint, Gyan Books Pvt. Ltd. (2013)
6. Meneghetti, U.: Elastodynamic analysis of an ancient mechanism. In: Proceedings of the TrISToMM 2015, pp. 303–309. ISBN: 978 605 84220 0 1 (2015)
7. Nicholson, J.: The operative mechanic (1826). Reprint, ISBN: 978933 3136938 (2013)
8. Rees, A.: The cyclopædia or universal dictionary of arts, sciences, and literature, vol. XXXVIII. Longman, Hurst, Rees, Orme and Brown, London (1819)
9. White, L. Jr: Medieval Technology & Social Change. Oxford University Press, ISBN: 978 0 19 500266 9 (1962)

An Analysis of Micro Scratches on Typical Southern Chinese Bronzes—A Case Study of Crawler-Pattern Nao and Chugong Ge

Lie Sun, Xiaowu Guan and Shilei Wu

Abstract Based on the investigation of typical southern Chinese bronze, the authors develop a method to analyze the manufacturing processes and using technique of ancient metallic wares. The Clay mold casting was one of the most important mechanical technologies in the Bronze Age of China, when the majority of bronzes were produced. The style and foundry techniques of bronzes found in southern China (about 1,600 BC–200 BC) are different from those of Shang dynasty (1,600 BC–1,046 BC). Thirty typical items were selected from hundreds of southern Chinese bronzes of the Hunan Provincial Museum ("HPM"), and analyzed with the stereoscopic microscope ("SM") via the "low-power magnification method" (20×–200×). Seven types of causes of micro scratches are observed as follows: molding, casting, dressing, use wear, surface intervention, rust and unidentified causes. Previously, the micro scratch analysis was seldom used in technological analysis of bronzes. For this paper, we collected clay mold assembly and separation information by using this method, more precisely distinguished the direct mold casting patterns from reversed mold casting patterns, and thus provided a more profound and detailed understanding of bronze foundry techniques.

Keywords Micro scratch · Clay mold casting · Casting flash · Mold casting pattern · Bronze

---

L. Sun (✉) · X. Guan · S. Wu
Institute for the History of Natural Sciences,
Chinese Academy of Sciences, Beijing, China
e-mail: sunlie@ihns.ac.cn

X. Guan
e-mail: gxiaowu@163.com

S. Wu
e-mail: wushilei@ihns.ac.cn

1 Introduction: Clay Mold Casting and Chinese Bronzes

The metallic material technology of ancient China approximately started from the mid to late Neolithic Age, i.e., Yangshao and Longshan cultures (c. 3,000 BC–c. 1,900 BC). Before Han dynasty (206 BC–220 AD), bronze metallurgy and processing are the most essential metallic technologies. According to archeological evidence, although bronze technologies in China emerged later than those on the Iranian plateau in the 5th millennium BC, and was likely influenced by Central Asia [1]. China's bronze technologies later maintained the tradition that took clay mold casting as the main foundry technique [2].

China's clay mold casting started from Longshan culture of the Neolithic Age and reached its peak in Shang (1,600 BC–1,046 BC) and Western Zhou (1,046 BC–771 BC) dynasties. From Eastern Zhou dynasty (770 BC–256 BC), although metal mold casting and investment casting already appeared, clay mold casting remained the most important foundry technology in the Bronze Age, and even extended into the Iron Age. With clay mold casting, most metallic farm implements, manipulative tools, ritual vessels, weapons, chariot fittings, harnesses and other mechanical components were manufactured (during the period) [3–5].

The techniques of clay mold casting could be divided into five steps in general.

(1) Selecting and processing molding materials. The materials were mainly made of clay with grains of sand, and some materials applied clay with grains of shells. Such mixed materials generally had to go through such processing procedures as elutriation, which not only could pick out appropriate granules, but also could increase the SiO_2 content and reduce the CaO and MgO contents in clay molds, thus improving the refractoriness of molds.

(2) Making clay models, clay molds and mold cores. Generally speaking, models were made first. Then, with reference to models, molds were worked out. Finally, molds were put together for casting. The compositions of the materials for the above-mentioned three parts were different. The former two contained more clay so as to improve plasticity and reproducibility, while the latter contained less clay with bigger grains of sand for the purpose of improving air permeability. After taking shape, a clay model needs to be fired into quasi-terracotta to fix the shape. The patterns in high relief might be formed with incremental layering, while refined patterns might be directly cast. After the casting process, direct mold casting patterns would appear on the bronze. A mold could be either directly engraved and shaped or copied from a model. After the casting process, reversed mold casting patterns corresponding to the patterns thus formed would appear on the bronze. There existed two ways of making a mold core. In most cases, a core was made by machining the model from its surface by one layer, the thickness of which was equivalent to that of the bronze. Meanwhile, in some cases, a mold core was made separately. The molding and pattern-making of a bronze were almost always the most labor-consuming procedure, and also the key procedure determining the success of the making. Despite the fact that direct mold casting patterns and

reversed mold casting patterns could help us understand specific foundry techniques, the differences between the two techniques are not always evident.

(3) Drying and baking of molds. Placing a mold in a cellar to dry could slowly evaporate the moisture in the clay mold, preventing the mold from cracking. Molds were already baked in Western Zhou dynasty for three purposes: 1. Removing moisture, 2. Decomposing carbonates and hence releasing part of air, and 3. Making molds reach a quasi-terracotta status for fixing the shape and increasing strength.

(4) Casting. Before mid-Shang dynasty, bronzes were dominated by whole casting. The birth of separate casting had a great impact on bronze technologies in China. With this method, the body of a vessel and its accessories were cast separately, and then joined together by casting. Many large bronzes with multiple structures and complicated patterns were made with this technique.

V. Post-casting processing. After casting, craftsmen not only had to remove the protruding parts in the inlet and the outlet of a bronze, but also need to polish its surface. Weapons often need to be sharpened.

(5) The time when bronzes were made in southern China is roughly at the same period of Shang and Zhou dynasties in northern China. In comparison, although both northerners and southerners mainly cast with clay molds, as southern Chinese bronzes are different from northern ones in terms of shape and appearance, pattern and specific technique [6, 7], they are important for us to understand the dissemination and evolution of bronze foundry technology [8, 9]. Nevertheless, past studies were mainly based on naked eye observations, which could only acquire limited information. In 2015, the Research Department on the History of Natural Sciences at the Chinese Academy of Sciences and the Hunan Provincial Museum ("HPM") launched a joint research program, which selected 30 representative items from hundreds of southern Chinese bronzes from the collection of the museum and analyzed the micro scratches on these bronzes by introducing the digital super depth-of-field ("DOF") stereoscopic microscope ("SM"). This article takes Crawler-pattern Nao, a kind of big cymbals, and Chugong Ge, a kind of dagger-axe, for example to illustrate the features of the technology and research method.

2 Analytical Method

The SM presents stereoscopic images with dual-channel optical paths, and can be used to observe the three-dimensional status of the samples. Because of the strong stereoscopic impression of its images, the large diameter of its field of view, and its great DOF, the SM can act as a key instrument to collect information of micro scratches on bronzes. When being operated, if the magnification power is limited between 20× and 200×, the microscope will be more suitable for collecting clear

Table 1 Classification of bronze micro scratches (BR-MS)

Code of type	Cause of micro scratches
BR-MS1	Molding: model, mold, core, core brace, gasket, mortise and tenon joint, supporting ornament, and mold assembly
BR-MS2	Casting: casting, supplementary casting, casting defects
BR-MS3	Dressing: cutting the pouring channel, polishing defects; post-casting processing, machining, polishing, mounting, grinding, painting, gilding, plating and forging
BR-MS4	Use wear
BR-MS5	Burying, handing down and later intervention
BR-MS6	Rust
BR-MS7	Unidentified causes

Instrument: Keyence VHX-1000 digital ultra DOF SM
Parameters: CCD physical pixel: 2110,000 valid pixels; real-time dynamic image collection; 20×–200× lens of DOF: 34–0.44 mm, scope of observation: 19.05–1.14 mm, working distance: 25.5 mm; 0×–50× lens of working distance: 95 mm
Time: June 11–13 and July 8, 2015
Venue: Storeroom of Cultural Relics, Official Kiln Base, HPM

and complete information of processing, use wear and rust. Therefore, this method could be dubbed as the "low-power magnification method" for studying micro scratches on bronzes.

In most cases, as sample preparation is not required for a bronze to be observed, the SM is well adapted to collecting in situ non-destructive information of micro scratches. Besides, the digital SM also boasts such functions as super DOF stacking, and saving and processing of digital images. However, limited to the sizes of bronzes and its stage, an SM is, sometimes, unable to achieve super DOF stacking, or unsuitable for observing oversized bronzes.

With reference to the causes and characteristics of the appearance and shape of the micro scratches on bronzes, micro scratch information can be classified into the following seven categories (Table 1).

3 Crawler-Pattern Nao (Big Cymbals, HPM Code: 20828)

The big cymbals handle is cylindrical with a broader bottom and a thinner top, while the section is not a regular circle (Fig. 1). The external diameter of the top of the big cymbals handle is 66.4–60 mm, while the inner diameter is 51.8–56 mm. The external diameter of the bottom is 66–75 mm. The drum on each side is decorated with a crawler in high relief in the shape of a stretched "S". The head of the crawler is triangular. The body of the crawler is sturdy, and the tail is thin and pointed. On each side of the body, there is a pair of feet presented with a thin protruding line. The lower part of one crawler has parallel hair. The overall height of the bronze is 415 mm, the height of the cymbals handle is 121 mm, the distance

Fig. 1 *Front view, side view* and *bottom view* of Crawler-pattern big cymbals

between the two drums is 310 mm (the interior distance is 281 mm). In our analysis, Side A is the one in which in the middle of the upper line of the front of the cymbals body exists a black breakage; the other side is Side B; and the interior is Side C.

There is an evident boundary trace along which the mold was assembled (Fig. 2). Besides, the mismatched scratches of the two sides of the loop are also evident. The pattern is not connected to the mold lines, which might be formed when the mold was put together (BR-MS1). And there exists disorderly linear scratches on the surface of the pattern on the loop.

Regarding the reversed jagged mold casting pattern, on its right-hand side (Fig. 3), there exist indentures, which were likely made in the course of mold making (BR-MS1). And there exist many casting pores on the surface (BR-MS2).

To its lower left, there was a concave cambered surface made by post-casting processing (BR-MS3) (Fig. 4). On the rust of the tip, there appear intervention linear scratches (BR-MS5) (Fig. 4). The cutting scratches on the exposed bronze base above the right corner of the cymbals to the right were likely produced when craftsmen removed rust in different directions (BR-MS5).

A hole formed by a casting defect (BR-MS2) (Fig. 5). Its typical rust lamination is presented (BR-MS6). And there are some scratches with unidentified causes appear on the surface (BR-MS7).

Fig. 2 A boundary trace above the loop to the *left* on Side A

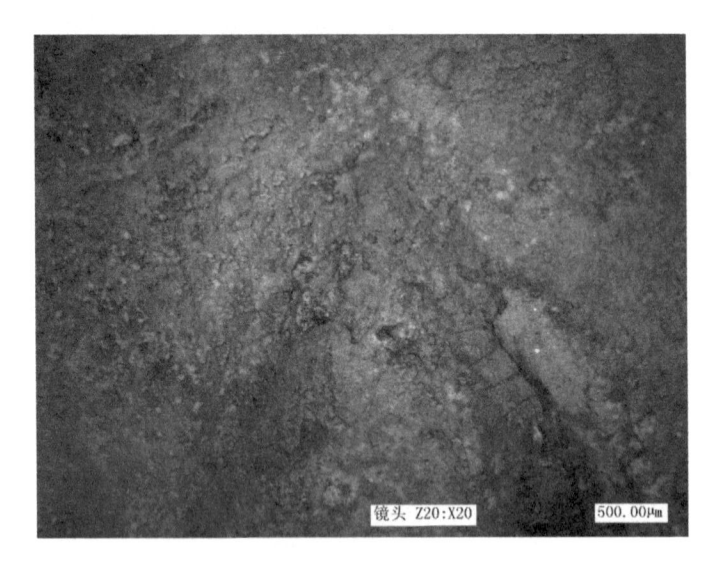

Fig. 3 Jagged pattern on the *left* body of Side B

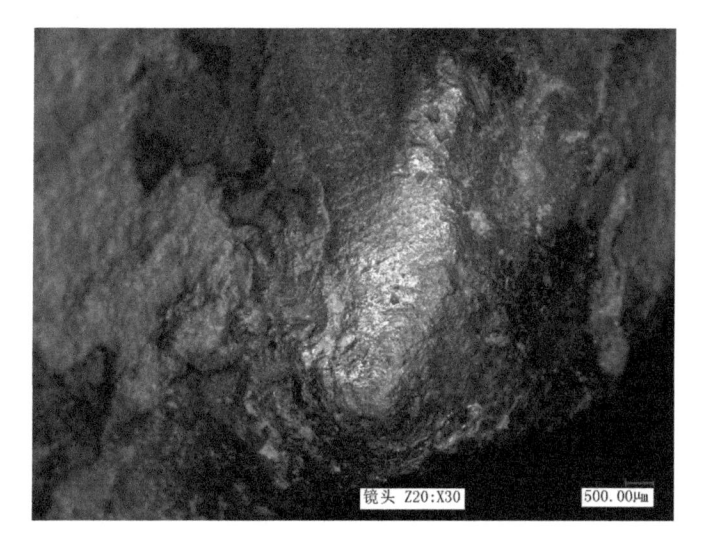

Fig. 4 The tip of the *right* corner of the corner on Side A

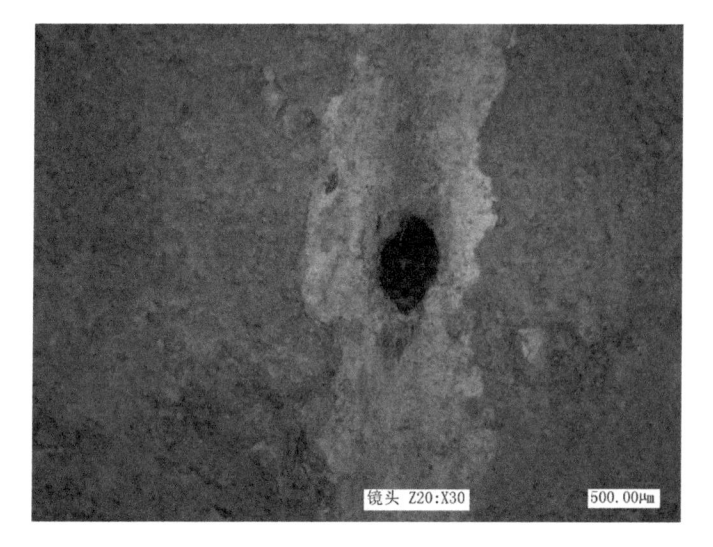

Fig. 5 The hole in the *right* body on Side A

Cutting scratches of later intervention are evident on the cross section (BR-MS5) (Fig. 6), exposing the bronze base. On the surface, a transparent sealing material was painted (BR-MS5).

It is evident that damages was resulted from casting defects appear on the surface (BR-MS2) (Fig. 7). There exists typical rust lamination (BR-MS6). And there are tiny round dents whose rims are circular (BR-MS7).

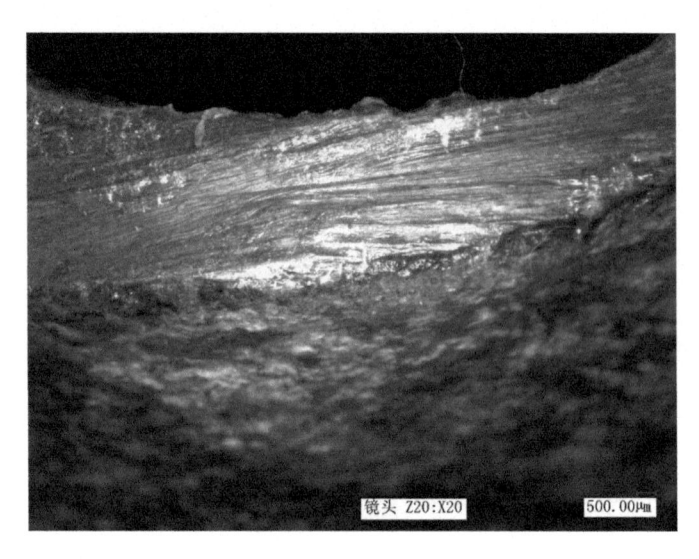

Fig. 6 Rim of the cymbals Handel on Side B

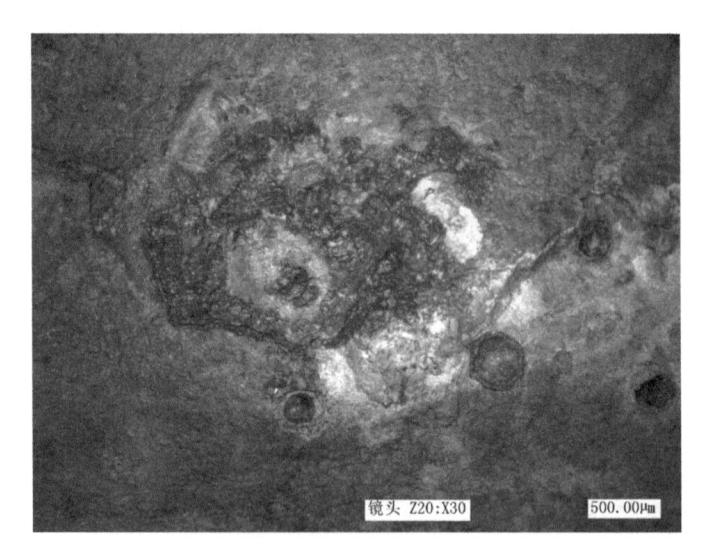

Fig. 7 Part of the crawler pattern of the left drum on Side A

4 Chugong Ge (Dagger-Axe, HPM Code: 39212)

The triangular dagger-axe is 212.9 mm long and 89.6 mm wide, has a date pit-shaped hole in the handle with its long axis in the axial direction of the dagger-axe. The round hole is almost round with diameters of 17.3 mm × 18.7 mm. The edge of the dagger-axe is slightly damaged (Fig. 8). The main body of the

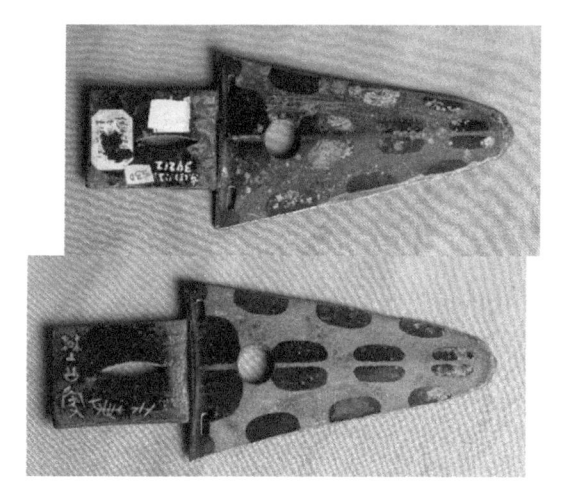

Fig. 8 The side without inscriptions and the side with inscriptions of Chugong Ge

bronze is covered with dense green rust, among which there are black shell-shaped spots. In our analysis, the side with no inscriptions is Side A, and the side with inscriptions is Side B.

Tracks of the mold assembly could be seen on the inner wall (BR-MS1) (Fig. 9). At the same time, dressing scratches can be seen on the edge (BR-MS3).

A casting burr line can be seen (BR-MS1) (Fig. 10); there exist evident shrinkage cavities (BR-MS2). There exist continuous scratches on the end faces of

Fig. 9 Part of the round hole on Side A

Fig. 10 Lateral sides of the handle on Side A

the upper and lower lateral sides of the handle, which might result from dressing, post-casting processing (BR-MS3) or use wear (BR-MS4).

A skin lamination is evident (BR-MS2) (Fig. 11). The intervention scratches appear on the edge (BR-MS3); and dressing scratches appear on the surface on black spots (BR-MS5).

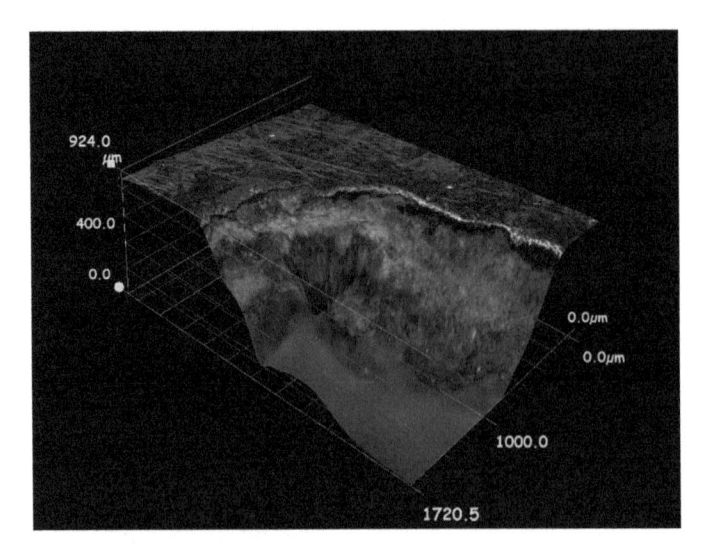

Fig. 11 Cross section of the hole on Side A

Later intervention and processing scratches can be seen on the rust at the rim (BR-MS5) (Fig. 12) and were likely produced when the dagger-axe was sharpened; and crossed and overlapped scratches also appear on the surface (BR-MS7).

Two protruding strips can be seen (BR-MS1) (Fig. 13). The crossed and overlapped scratches resulted from later intervention can be seen on the surface of and

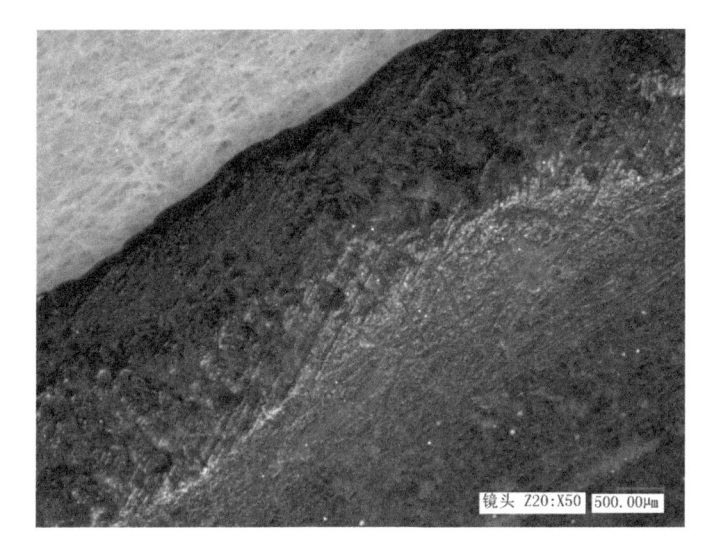

Fig. 12 Rim of the edge of Side A

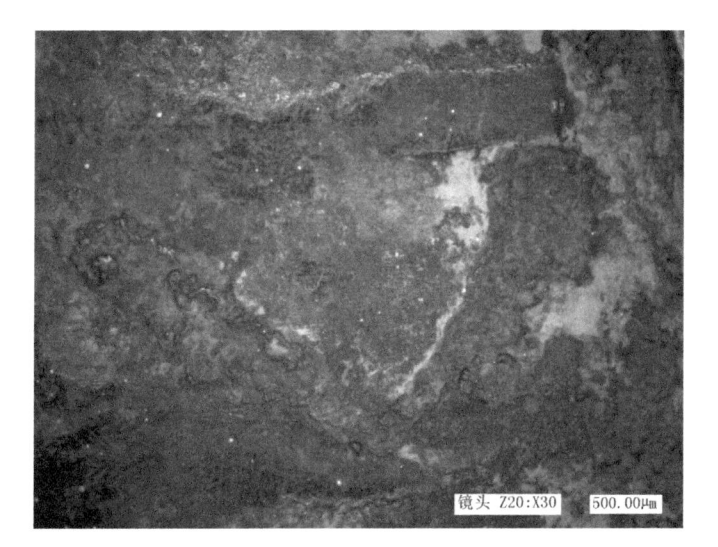

Fig. 13 Part of the surface close to the handguard on Side A

between the two protrusions (BR-MS5). And between the two protrusions, there exists a piece of "V"-shaped rust (BR-MS6).

Three tiers of outlines can be seen and might be connected to use wear (BR-MS4) (Fig. 14). A fracture appears on the tip and might results from use wear (BR-MS4). Rust (BR-MS6) presents different layers with distinct color differences. Vertical and horizontal scratches that are suspected to derive from later intervention can also be seen on different layers of rust (BR-MS5).

Two fractures appear that one deep and one shallow (Fig. 15). The two fractures expose the bronze base. The two fractures are suspected to be use wear (BR-MS4). Parallel scratches resulted from later modification can be found on the rim of the edge, the rust and the black spots (BR-MS5). There is a deep "Y"-shaped scratch and a shallow scratch starting from the rim of a black spot and extending from lower right to upper left can be observed (BR-MS5). And the green rust and the black spot rust (BR-MS6) differ from one another distinctly in terms of color.

With reference to the "U"-shaped cross section of the shape (Fig. 16), we can see the outline and strokes of character "Ge". The character presents a typical casting features (BR-MS1). There are pores and casting defects around character "Ge" (BR-MS2). As the boundary between black spots and character "Ge" is clear, the inscription should be cast first, while black spots should formed later (BR-MS3). And there are slanting scratches resulting from later intervention under character "Ge" (BR-MS5).

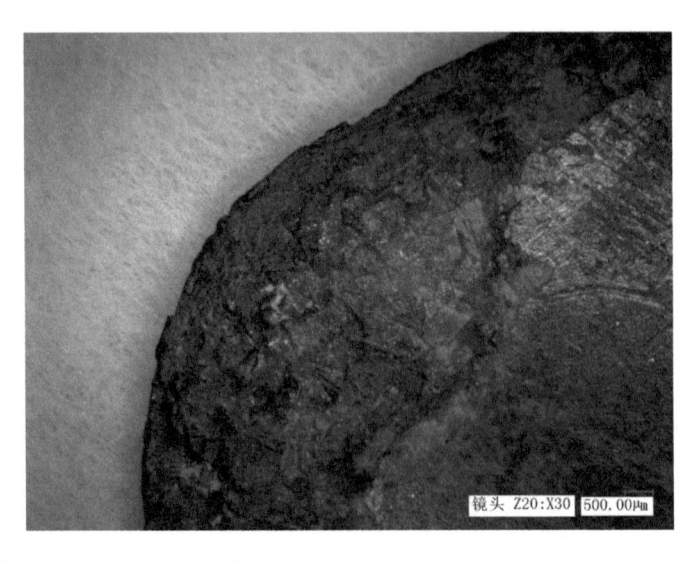

Fig. 14 Tip of the dagger-axe on Side A

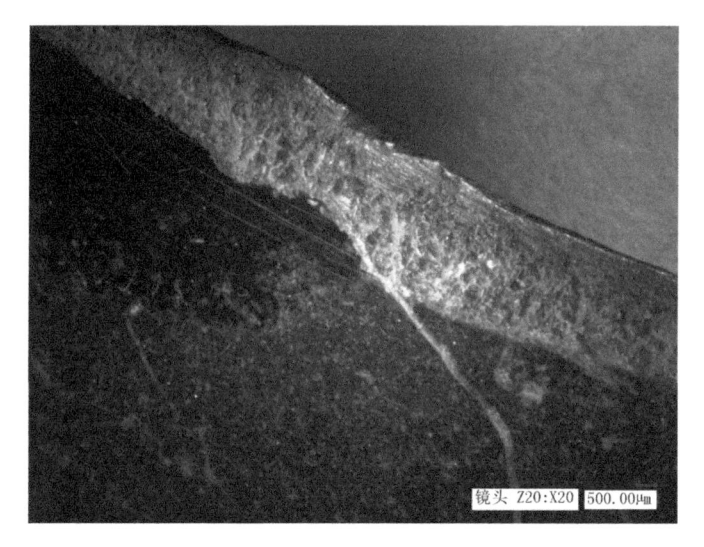

Fig. 15 Rim of the edge of Side A

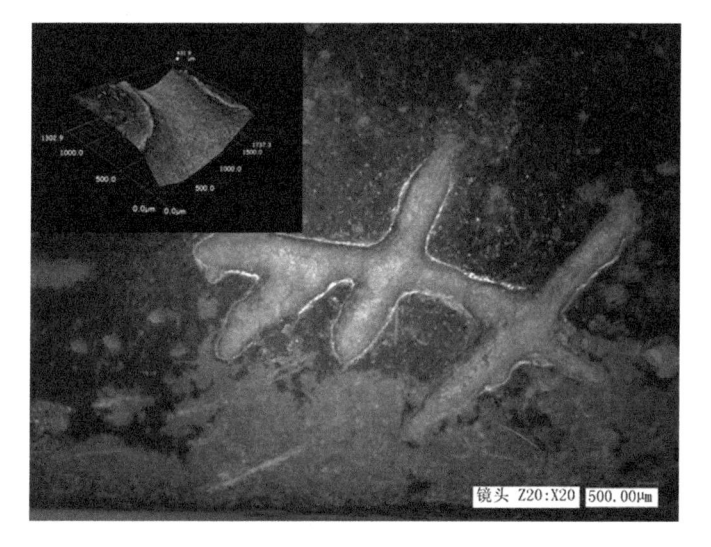

Fig. 16 Inscription "Ge" and partial DOF stacking (the *upper left* image)

5 Conclusions

(1) Stereoscopic microscope (SM) can collect information of micro scratches resulting from molding, casting, dressing, post-casting processing, use wear, burying, handing down, later intervention, rust and unidentified causes relating to bronzes. Compared with previous technological studies of bronzes, this

project explores the method of observing micro scratches through the SE, and realizes the irreplaceable role played by the SM in micro scratch analysis, technique study and preliminary research for cultural relics protection/preservation.
(2) The seven types of causes of micro scratches both involve technical foundry processes, such as pre-casting treatment, casting and post-casting treatment, and are closely connected to use wear, rust, later intervention and damage. Through the "low-power magnification method", we could observe the above scratches or traces on almost all items. Among others, certain scratches that can hardly be observed by using the macroscopic observation method can also be clearly and distinctly presented, while through the method, we could acquire a huge amount of information. In particular, as the information collected is typical, and hence be served as more reliance evidence for further technological studies.
(3) Ritual musical instruments and weapons played an important role among Chinese bronzes in antiquity. Technically speaking, casting bronzes in an assembled clay mold with split-mold parts was a dominant method [10]. The abundant information presented by the huge and heavy Big cymbals and the small and thin Ge under the microscope demonstrates the maturity and complicity of the molding and casting technology of southern Chinese bronzes during about 3000 BC.
(4) Although it is not conventianl, the method for researching ancient bronze with micro scratch analysis of SM is not exchangeable or substitutable. For our next step, it is expected to be used with other methods, such as the macroscopic observation method, the silica gel impression method, the scanning digital microscope and simulation experiments, so that the benefits of different methods can be fully exploited when micro scratches on bronzes of different sizes, types and storage conditions are studied and that more sophisticated and comprehensive issues could be resolved.

Acknowledgments We gratefully acknowledge the support of the Hunan Provincial Museum, and the Center of Scientific Studies on Cultural Heritage at Chinese Academy of Sciences, which made relevant research for this paper possible. We deeply obliged to Professor Rongyu Su for his encouragement and generosity to share with us his systematic research concerning the bronzes discussed above. We are grateful to professor Juliang Fu, Liang Liu, Yuan Li, Xiaoyan Wu, Xin Yuan, Wenli Zhou and our other team members of the project "Technological Research on Typical Bronzes of HPM". We thank professor Dongpo Hu, Tingrui Zhao, Xingang Tan and Zhongtian Li for their assistance at various stages of this project.

References

1. Metropolitan Museum of Art: Treasures from the Bronze Age of China: an Exhibition from the People's Republic of China, pp. 11–12. Metropolitan Museum of Art, New York (1980)
2. Lu, J.Y., Hua, J.M.: History of Science and Technology in China: Machinery, p. 177. Science Press, Beijing (2000)

3. Pigott, V.C.: The Archaeometallurgy of the Asian Old World. UPenn Museum of Archaeology, pp. 126–127 (1999)
4. Lu and Hua, p. 177
5. Lu, Y.X.: A History of Chinese Science and Technology, pp. 521–522. Springer, Heidelberg (2014)
6. Kane, V.C.: The independent bronze industries in the South of China contemporary with the Shang and Western Chou dynasties. Arch. Asian Art **28**, 77–107 (1974)
7. Ren, W.: Bronzes in the drainage basin of the Yangtze River and archeology of Shang dynasty. Relics South **26**(2), 31–48 (1996)
8. Shi, J.S.: A study of bronzes in the drainage basin of the Yangtze River, p. 1. Cultural Relics Press, Beijing (2003)
9. Xiang, T.C.: A summary of bronzes of Shang and Zhou dynasties in the drainage basin of the Xiang River. Hunan Univ J (Social Science Edition) **21**(5), 42–47 (2007)
10. Su, R.Y., Hua, J.M., Li, K.M.: Chinese Metallic Technology in Antiquity, p. 87. Shandong Science and Technology Press, Jinan (1995)

Printed by Printforce, the Netherlands